ıe-Added
ˈ By-Product

Recovery of High Value-Added Compounds from Food By-Product

Guest Editors

Remedios Yáñez
Beatriz Gullon

MDPI

Basel • Beijing • Wuhan • Barcelona • Belgrade • Novi Sad • Cluj • Manchester

Guest Editors

Remedios Yáñez
Department of
Chemical Engineering
University of Vigo
(Campus Ourense)
Ourense
Spain

Beatriz Gullon
Department of
Chemical Engineering
University of Vigo
(Campus Ourense)
Ourense
Spain

Editorial Office
MDPI AG
Grosspeteranlage 5
4052 Basel, Switzerland

This is a reprint of the Special Issue, published open access by the journal *Foods* (ISSN 2304-8158), freely accessible at: https://www.mdpi.com/journal/foods/special_issues/Food_By-Product.

For citation purposes, cite each article independently as indicated on the article page online and as indicated below:

Lastname, A.A.; Lastname, B.B. Article Title. *Journal Name* **Year**, *Volume Number*, Page Range.

ISBN 978-3-7258-7641-9 (Hbk)
ISBN 978-3-7258-7642-6 (PDF)
https://doi.org/10.3390/books978-3-7258-7642-6

Contents

foods MDPI

ditorial

Recovery of High Value-Added Compounds from Food By-Product

Beatriz Gullón [1,*] and Remedios Yáñez [1,2,*]

[1] Department of Chemical Engineering, Faculty of Science, University of Vigo (Campus Ourense), As Lagoas, 32004 Ourense, Spain
[2] CINBIO—Centro de Investigacións Biomédicas, Universidade de Vigo, 36310 Vigo, Spain
* Correspondence: bgullon@uvigo.es (B.G.); reme@uvigo.es (R.Y.)

The agri-food industry generates large quantities of by-products, both of animal and vegetable origin, which are currently discarded or destined to low-value-added applications [1]. However, they frequently present very interesting chemical compounds with great application potential in the pharmaceutical, food, or cosmetics industries. Therefore, to move towards a circular economy with a high level of resource efficiency, it would be necessary to develop integrated multi-product biorefineries based on environmentally friendly techniques that extract the most value from these by-products, a strategy that would minimize adverse effects on human health and the environment while improving the economic competitiveness of these sectors [2].

In this special issue, interesting valorization opportunities for several agri-food by-products are explored in 9 papers, by-products such as fruit wastes like melon peel [3], soybean [4] and chestnut [5] residues, alfafa [6], and fish [7] and meat remains [8]. Authors evaluate different green technologies for the recovery of several valuable fractions, such as proteins, carbohydrates, pectic oligosacharides, and phenolic compounds, and assess their potential as sustainable sources of bioactive compounds, analysing their antioxidant, antibacterial, anti-tyrosinase, and anti-inflammatory activities as well as their prebiotic potential. On the other hand, authors review the biological activities exerted by tannins extracted from by-products of the agri-food industry [9], the employment of innovative techniques involved in the conversion of rice bran into valuable food ingredients [10] and the application of agro-industrial wastes in aquaculture [11].

Author Contributions: B.G. and R.Y. conceived and wrote this editorial. All authors have read and agreed to the published version of the manuscript.

Funding: This research received no external funding.

Conflicts of Interest: The authors declare no conflict of interest.

References

1. Gullón, P.; Gullón, B.; Romaní, A.; Rocchetti, G.; Lorenzo, J.M. Smart advanced solvents for bioactive compounds recovery from agri-food by-products: A review. *Trends Food Sci. Technol.* **2020**, *101*, 182–197. [CrossRef]
2. Del Castillo-Llamosas, A.; del Río, P.G.; Pérez-Pérez, A.; Yáñez, R.; Garrote, G.; Gullón, B. Recent advances to recover value-added compounds from avocado by-products following a biorefinery approach. *Curr. Opin. Green Sustain. Chem.* **2020**, *28*, 100433. [CrossRef]
3. Rico, X.; Gullón, B.; Yáñez, R. Environmentally friendly hydrothermal processing of melon by-products for the recovery of bioactive pectic-oligosaccharides. *Foods* **2020**, *9*, 1702. [CrossRef] [PubMed]
4. Le, B.; Pham, T.N.A.; Yang, S.H. Prebiotic potential and anti-inflammatory activity of soluble polysaccharides obtained from soybean residue. *Foods* **2020**, *9*, 1808. [CrossRef] [PubMed]
5. Tsurunaga, Y.; Takahashi, T. Evaluation of the antioxidant activity, deodorizing effect, and antibacterial activity of 'Porotan' chestnut by-products and establishment of a compound paper. *Foods* **2021**, *10*, 1141. [CrossRef] [PubMed]
6. Solarte, D.A.; Ruiz-Matute, A.I.; Chito-Trujillo, D.M.; Rada-Mendoza, M.; Sanz, M.L. Microwave assisted extraction of bioactive carbohydrates from different morphological parts of alfalfa (*Medicago sativa* L.). *Foods* **2021**, *10*, 346. [CrossRef]

7. Steinsholm, S.; Oterhals, Å.; Underhaug, J.; Aspevik, T. Emulsion and surface-active properties of fish soluble based on dire extraction and after hydrolysis of Atlantic cod and Atlantic salmon backbones. *Foods* **2021**, *10*, 38. [CrossRef] [PubMed]
8. Tedeschi, T.; Anzani, C.; Ferri, M.; Marzocchi, S.; Caboni, M.F.; Monari, S.; Tasson, A. Enzymatic digestion of calf fleshing me by-products: Antioxidant and anti-tyrosinase activity of protein hydrolysates, and identification of fatty acids. *Foods* **2021**, *10*, 75 [CrossRef] [PubMed]
9. Fraga-Corral, M.; Otero, P.; Echave, J.; García-Oliveira, P.; Carpena, M.; Jarboui, A.; Nuñez-Estevez, B.; Simal-Gandara, J.; Priet M.A. By-products of agri-food industry as tannin-rich sources: A review of tannins' biological activities and their potential fc valorization. *Foods* **2021**, *10*, 137. [CrossRef] [PubMed]
10. Spaggiari, M.; Dall'Asta, C.; Galaverna, G.; del Castillo Bilbao, M.D. Rice bran by-product: From valorization strategies t nutritional perspectives. *Foods* **2021**, *10*, 85. [CrossRef] [PubMed]
11. Leyva-López, N.; Lizárraga-Velázquez, C.E.; Hernández, C.; Sánchez-Gutiérrez, E.Y. Exploitation of agro-industrial waste a potential source of bioactive compounds for aquaculture. *Foods* **2020**, *9*, 843. [CrossRef] [PubMed]

foods

MDPI

Article

Evaluation of the Antioxidant Activity, Deodorizing Effect, and Antibacterial Activity of 'Porotan' Chestnut By-Products and Establishment of a Compound Paper

Yoko Tsurunaga * and Tetsuya Takahashi

Faculty of Human Science, Shimane University, Shimane 690-0823, Japan; takahashi@hmn.shimane-u.ac.jp
* Correspondence: ytsurunaga@hmn.shimane-u.ac.jp; Tel.: +81-852-32-9806

Abstract: Chestnuts are widely cultivated for their edible portion (kernel), whereas the non-edible parts are discarded. To enable the utilization of the by-products of processed chestnuts, we separated them into green and brown burs, shells, inner skin, and leaves, and analyzed the bioactive properties of the ground components. We also created a composite paper, comprising the inner skin, and examined its deodorant properties. It was revealed that the inner skin had the highest functionality and showed potent antioxidant, antibacterial, and deodorant properties. Furthermore, when we produced a paper, containing 60% inner skin, and examined its deodorant properties, we found that it was highly effective in deodorizing ammonia and acetic acid gases. These results show that the inner skin of chestnuts is a promising material for developing hygiene and other products.

Keywords: 'Porotan' chestnut; *Castanea crenata* Sieb. et Zucc; antioxidant activity; deodorant activity; antibacterial activity; compound paper

1. Introduction

Chestnut cultivation and processing are performed worldwide. Chestnuts cultivated worldwide include the Japanese chestnut (*Castanea crenata* Sieb. et Zucc.), Chinese chestnut (*Castanea mollissima* Blume), and European chestnut (*C. sativa* Mill.) cultivated in Southern Europe and Asia Minor. According to the data from the Food and Agriculture Organization, Asia is the leading global producer of chestnuts (86.5%), followed by Europe (9.6%) and the United States (3.8%) [1]. The kernel (edible part) of chestnuts is used to prepare glacé, pastes, and syrups [2]; however, additional parts of the chestnut (*Castanea crenata* Sieb. et Zucc.), including the leaves, burs, shells, and the inner skin, are currently not used. Chestnut shells contain various phenolic compounds [3], and the tough skin contains a large amount of tannins [4], which show antioxidant [3,5,6], fat absorption-inhibiting [7], and anti-adipogenic activities. Tannins also reportedly have preventive effects against heart disease [8], diabetes [9], and cancer [10], as well as antibacterial [11,12] and deodorizing properties [13]; therefore, the utilization of fruit peels is considered important from an industrial perspective [14]. However, the existing Japanese varieties of chestnuts are difficult to peel, and the bioactivity of the inner bark of trees have not been investigated. The shell and inner skin of the variety 'Porotan', cultivated in Japan, can be peeled extremely well [15,16]; therefore, the inner skin can be used as a by-product after the processing of chestnut kernels. In this study, we aimed to utilize the unused parts of chestnuts for food and industrial applications and analyzed the functionality (polyphenol content, antioxidant activity, anti-bacterial activity and deodorizing activity) of the chestnut burs (green and brown), leaves, shells, inner skin, and kernel. Secondly, we tested the functionality (deodorizing activity) of the produced compound paper, which contains 60% inner skin. It was clarified that the inner skin had the highest polyphenol content, antioxidant, anti-bacterial, and deodorant properties. Further, we found that the paper containing 60% inner skin is highly effective in deodorizing ammonia and acetic acid gases. This research will lead to the effective utilization of the large amounts of by-products generated from chestnut cultivation and processing.

2. Materials and Methods

2.1. Samples

2.1.1. Japanese Chestnut 'Porotan'

Burs (green and brown), shells, inner skin, kernel, and leaves were obtained from Japanese chestnut 'Porotan' trees (scientific name: *Castanea crenata* Sieb. et Zucc.) in the Institute of Fruit Tree Science of Ibaraki Prefecture, Japan, during September 2017 (Figure 1). The materials were freeze-dried and ground into a powder for subsequent analyses. The powder was ground in a blender for approximately 120 s, and then sieved through a mesh sieve (1 mm) to obtain samples.

Figure 1. Photographs of Japanese chestnut 'Porotan' components. (**a**): Burs (green), (**b**): burs (brown), (**c**): shells, (**d**): inner skin, (**e**): kernel, (**f**): leaf.

2.1.2. Effects of Chestnut Samples on Bacteria

S. aureus NBRC 12732 and *E. coli* NBRC 3301 were obtained from the National Institute of Technology and Evaluation, Osaka, Japan, and used to analyze the antibacterial activities of leaves, burs (green and brown), shells, and inner skin in 'Porotan' trees.

2.1.3. Preparation of Compound Papers

Compound papers were prepared using a previously reported method [17–20].

2.2. Analysis of Properties

2.2.1. Extraction Method for Total Polyphenol Content (TPC), 2,2-Diphenyl-1-Picrylhydrazyl (DPPH), and Hydrophilic Oxygen Radical Absorbance Capacity (H-ORAC) Assays

The prepared powdered samples (200 mg) were added to 60% (v/v) ethanol aqueous (10 mL) and shook for 2 h at 40 °C. The extracted was filtered through a 0.45 μm filter.

2.2.2. TPC

The soluble polyphenol content of the extracts was measured by the Folin–Ciocalteu method [21]. The TPC was expressed as mg equivalent/100 g of dry matter, using catechin (CTN) as the standard (mg CTN eq/100 g DW).

2.2.3. DPPH Radical Scavenging Activity Assay

The antioxidant activity of the extracts was analyzed using the DPPH radical scavenging assay [22]. The DPPH value was expressed as μmol Trolox equivalent/g of dry matter (μmol TE/g DW).

2.2.4. Hydrophilic-ORAC (H-ORAC)

The antioxidant activity of the extracts was analyzed using the H-ORAC method [20,23]. H-ORAC values were expressed as Trolox equivalents of dry matter (μmol TE/g DW).

2.2.5. Antibacterial Properties

Antibacterial activity assays were performed by modifying methods by which we had previously determined the antibacterial activity [17–20,24]. Using these methods, the activity of each part of Porotan was examined, and all parts except for kernel showed very strong activity (a kill rate of 100%) with no differences. To clarify the difference in antibacterial properties in each part of Porotan, the experiment was performed by applying stricter conditions to the previous method [17–20,24]. Specifically, we increased the amount of a bacterial suspension from 100 to 200 μL, the sample amount was reduced from 200 to 40 mg. The analyzed parts included the burs (green and brown), shells, inner skin, and leaves, based on the results of antioxidant activities. For comparison, green tea, which reportedly contains potent antibacterial properties was used.

2.2.6. Evaluation of Deodorant Activity

Deodorant activity assays were performed by the modifying methods that we had previously determined [19,20,25]. The odor residual ratio was plotted against time. The odor residual ratio of the sample was calculated using Equation (1) [19,20,25].

$$\text{Odor residual ratio (\%)} = \frac{\text{Measured gas concentration}}{\text{Initial concentration}} \times 100 \quad (1)$$

2.2.7. Scanning Electron Microscopy

The fixed samples were dried at room temperature for 24 h, and then gold was vapor-deposited on the sample using an ion sputterer (E-1010, Hitachi High-Technologies Corp., Tokyo, Japan). The surface of the inner skin of 'Porotan' chestnuts was observed using a scanning electron microscope S-3000N (Hitachi Science Systems, Ltd., Tokyo, Japan). The microscope was operated at an acceleration voltage of 20 kV.

2.2.8. Statistical Analysis

Data from TPC, DPPH, and H-ORAC assays were statistically analyzed using the SPSS statistical analysis software (version 25.0, SPSS Inc., Chicago, IL, USA). Results are expressed as the mean ± standard error. Data were tested using one-way-ANOVA, followed by Tukey's test for multiple comparisons.

3. Results and Discussion

3.1. TPC and Antioxidant Activity

Figure 2 shows the TPC and antioxidant activity of each of the following parts of the Porotan: burs (green and brown), shells, inner skin, and leaves. The TPCs in the green burs, brown burs, shells, inner skin, kernels, and leaves were 8770 ± 235, 6047 ± 125, 3325 ± 110, 27,208 ± 110, 12 ± 4.5, and 7414 ± 161 mg CTN eq/100 g DW, respectively. The antioxidant activities (DPPH) of the green burs, brown burs, shells, inner skin, kernels, and leaves were 653 ± 9.5, 384 ± 4.1, 269 ± 12.0, 2452 ± 45.0, 7.0 ± 1.0, and 461 ± 7.4 μmol TE/g DW, respectively. The antioxidant activities (H-ORAC) of the green burs, brown burs, shells, inner skin, kernels, and leaves were 833 ± 99.3, 55.3 ± 0.3, 304.3 ± 12.5, 1733 ± 91.2, 50 ± 0.8, and 1017 ± 16.0 μmol TE/g DW, respectively. The TPC and antioxidant activities (H-ORAC and DPPH method) of all parts of the chestnut, except for the kernel, were very high; the inner skin showed the highest TPC and antioxidant activity. Tuyen et al. [26] evaluated the antioxidant activity of the Japanese chestnut (barks, flowers, inner skin, kernels, and leaves) and reported that the inner skin showed the highest antioxidant activity in all the assays used (DPPH, 2,2-azinobis [3-ethylbenzothiazoline-6-sulfonic acid] diammonium salt, reducing power, and β-carotene bleaching methods). These findings are consistent with our results, which showed that the inner skin had the highest DPPH radical scavenging activity. In addition, we revealed that the inner skin had a high level of antioxidant activity with the H-ORAC method, which offers several advantages over the DPPH assays. The complete hydrogen atom transfer radical quenching mechanism presents a contrast and comparison to the electron transfer in the DPPH assays [27]. The ORAC assay uses peroxyl radicals that present a better model of the antioxidation reaction between oxidative lipids and reactive oxygen species [27]. The high values obtained via the ORAC assay are of great importance in terms of health functionality. Chestnut shells contain high concentrations of ellagic acid, a phenolic compound [5]. Ellagic acid is a dimeric derivative of gallic acid [28,29] and has certain health functions with specific biological effects, such as anti-angiogenesis [30], anti-cancer [10], anti-diabetes [31], and antioxidant [32]. Chestnut shells are expected to be used as highly valuable materials in the future. Additionally, we also found strong activity for burs, which Tuyen et al. [26] did not measure. The burs are a part of the plant that is removed during harvesting rather than peeling, thereby making it feasible to collect. In this study, we found that the green burs had higher TPC, DPPH, and H-ORAC values than the brown burs. Notably, the green burs had an H-ORAC value of 833 μmol TE/g DW, whereas that of the brown burs was 55.3 μmol TE/g DW, which was considerably low. Therefore, the inner skin, leaves, and burs (green) may present promising materials with antioxidant activity.

3.2. Antibacterial Activity

Gram-positive *S. aureus* and Gram-negative *E. coli* were used to assess the antibacterial activity of each part of the chestnut. The results are shown in Table 1. As a result, the growth values of both bacterial species yielded were above the JIS standard value for the assay of 0.5. Therefore, both of the species showed growth without inhibition, in the antibacterial assays. *S. aureus* showed growth from a viable count of 1.00×10^5 CFU/mL before incubation, to 1.04×10^{10} CFU/mL after 18 h of incubation in the culture medium without chestnut samples (control); therefore, the growth was increased by approximately 10^5-fold. In contrast, when cultured with the inner skin of the chestnuts, the viable count increased to 2.04×10^5 CFU/mL after incubation. Considering the viable count of 1.04×10^{10} CFU/mL of the control, the antibacterial effect corresponded to an approximately 10^5-fold decrease in cell count. In this experiment, we modified our previous methods, the amount of bacterial cell suspension was increased from 100 to 200 μL [17–20,24], the sample amount was reduced from 200 to 40 mg [17–19,24], and the experiment was conducted under conditions that enhanced bacterial growth. Therefore, the cultures incubated with green tea [33], which has high antibacterial activity, produced a bacterial count of 1.80×10^9 CFU/mL. This indicated that the inner skin might be highly effective in inhibiting *S. aureus* growth. The cultures incubated with green burs, brown burs, shells, and leaves showed a viable count of 3.84×10^6, 5.80×10^6, 1.36×10^6, and 1.16×10^6 CFU/mL, respectively, and thus, showed higher antibacterial activity than green tea. *E. coli* showed growth from a viable count of 1.00×10^5 CFU/mL before incubation to 2.80×10^9 CFU/mL after 18 h of incubation without chestnut samples (control); therefore, the growth was increased by approximately 10^5-fold. When the cultures were incubated with brown burs, the viable count increased to 2.28×10^7 CFU/mL after incubation. Considering the viable count of 2.8×10^9 CFU/mL of the control, the antibacterial effect corresponded to an approximately 100-fold decrease in cell count. Similar to the experiment using *S. aureus*, the amount of the bacterial cell suspension was increased from 100 to 200 μL, the sample amount was reduced from 200 to 40 mg, and conditions that enhanced bacterial growth were used. Therefore, the cultures incubated with green tea [33] produced a bacterial count of 4.64×10^8 CFU/mL. The cultures incubated with green burs, brown burs, shells, inner skin and leaves showed a viable count of 4.40×10^8, 2.28×10^7, 2.28×10^{10}, 2.16×10^8, and 1.24×10^9 CFU/mL, respectively. Only the brown burs had an antibacterial activity value of over 2, which meant that it reduced the number of cells by a factor of 100, and showed a higher antibacterial effect than green tea. This indicated that brown burs might be highly effective in inhibiting *E. coli* growth. *S. aureus* is a Gram-positive bacterium that contains a thick cell wall layer, mostly consisting of peptidoglycan, whereas the cell wall of the Gram-negative *E. coli* contains a small quantity of peptidoglycan. Generally, Gram-positive bacteria contain a peptidoglycan layer as thick as 20–80 nm, while Gram-negative bacteria contain a layer that is 7–8 nm thick; therefore, Gram-positive bacteria have a considerably thicker cell wall than that of Gram-negative bacteria. The DW of the bacterial cell wall is approximately 10% in Gram-negative bacteria. By contrast, the Gram-positive bacteria weigh as much as 90%. Based on the above results, it was predicted that *E. coli*, a Gram-negative bacterium, would show a higher inhibition of growth than *S. aureus* when treated with chestnut components; however, a different result was obtained. It has been reported that tea polyphenols show a higher antibacterial activity against *S. aureus* than against *E. coli* [33–35]; therefore, the result may not be explained by the thickness of the cell wall alone. The high antimicrobial activity of tea is due to the existence of catechins and polyphenols that damage the cell membranes of bacteria [36]. Catechins are more active against Gram-positive bacteria than Gram-negative bacteria [35]. EGCg inhibits cell membrane function by eliciting the leakage of small molecules from the intraliposomal space [35]. Porotan, like green tea, has a bacterial membrane-damaging effect, and this effect was more pronounced in *S. aureus* than in *E. coli*. JIS-L-1902 [37] specifies that an assay value of antibacterial activity of 2.0 or higher (a kill ratio of 99% or higher) indicates an antibacterial effect. The antibacterial activity of green and brown burs, shells, inner skin, and leaves against *S. aureus*, and that of brown burs against *E. coli*, met the criteria, indicating that they might act as potent antimicrobial materials. Each part of 'Porotan' was found to have a stronger antibacterial effect against *S. aureus* than against *E. coli*.

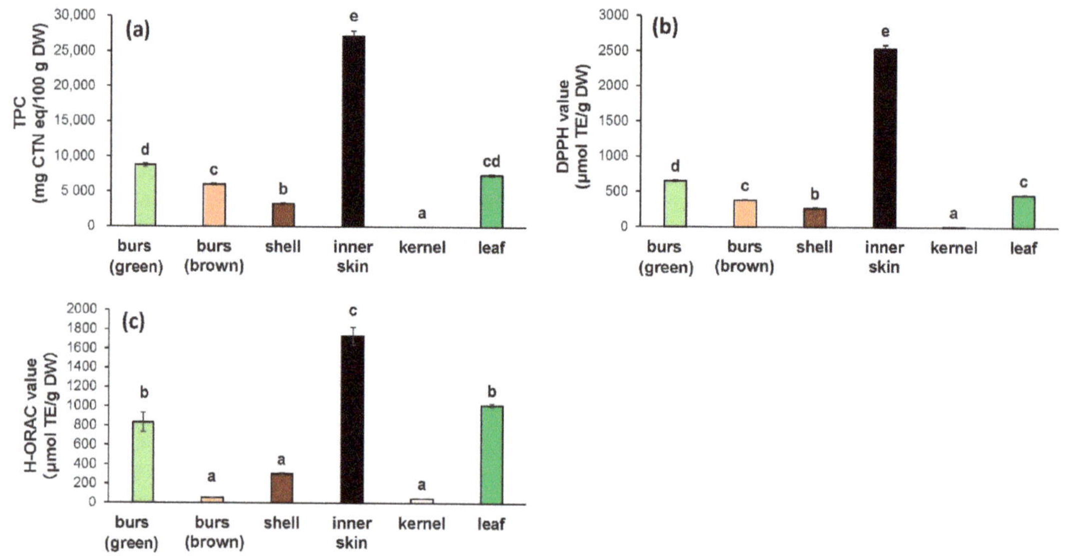

Figure 2. Antioxidant activity of 'Porotan' chestnut components. Antioxidant activity was evaluated as the TPC (**a**), DPPH radical scavenging activity (**b**), and H-ORAC (**c**). TPC, total polyphenol content; DPPH, 2,2-diphenyl-1-picrylhydrazyl radical scavenging activity; H-ORAC, hydrophilic oxygen radical absorbance capacity. Means are shown, with vertical bars indicating standard error ($n = 6$). The different letters (a–e) indicate statistical differences ($p < 0.05$).

Table 1. Antibacterial properties of 'Porotan' chestnuts against different bacteria.

Species of Bacterium	Sample	Incubation Time (h)	Antibacterial Properties		Bacteria Yielded Growth Values *3
			Viable Bacteria (CFU *1/mL)	Antibacterial Activity Value *2	
Staphylococcus aureus	Before incubation	0	1.00×10^5	–	
	Burs (green)	18	3.84×10^6	3.43	
	Burs (brown)	18	5.80×10^6	3.25	
	Shell	18	1.36×10^6	3.21	
	Inner skin	18	2.04×10^5	4.03	
	Leaf	18	1.16×10^6	3.28	
	Green tea	18	1.80×10^9	0.76	
	Only bacteria	18	1.04×10^{10}	–	5.02
Escherichia coli	Before incubation	0	1.00×10^5	–	
	Burs (green)	18	4.40×10^8	0.80	
	Burs (brown)	18	2.28×10^7	2.09	
	Shell	18	2.28×10^{10}	−0.91	
	Inner skin	18	2.16×10^8	1.11	
	Leaf	18	1.24×10^9	0.35	
	Green tea	18	4.64×10^8	0.78	
	Only bacteria	18	2.80×10^9	–	4.45

*1: Colony-forming units. *2: Antibacterial activity value = $(\log C_t - \log C_0) - (\log T_t - \log T_0)$. logCt: log (Ave-age (count of viable cell after incubation of bacteria alone for a given period)); $\log C_0$: log (average (count of viable cell before incubation)); $\log T_t$: log (average (count of viable cell after incubation with chestnut samples for a given period)); $\log T_0$: log (average (count of viable cell of the samples immediately after inoculation)). *3: Bacteria yield growth values = $\log C_t - \log C_0$.

3.3. Deodorizing Activity of the Inner Chestnut Skin and Compound Paper

The inner skin, which indicated high antioxidant and antibacterial activities, was selected for analysis of the deodorizing activity against various kinds of odors. The odorous gases used in the deodorization assays were ammonia (a basic odor), acetic acid (acidic odor), and acetaldehyde (neutral odor). We additionally analyzed the deodorizing properties of a compound paper containing 60% of inner chestnut skin (Figure 3). The change in odor residual ratio was found to be time-dependent (Figure 4). The results revealed that the inner skin and the compound paper effectively deodorized the basic odor of ammonia and the acidic odor of acetic acid. The odor residual ratio of ammonia gas decreased to 0% in just 10 min. The potent activity should be evaluated in further studies. The mechanism of the deodorization of ammonia gas by the inner skin and compound paper was considered to be that the hydroxyl groups of the polyphenols in the inner skin form hydrogen bonds with the ammonia molecules. In summary, certain OH groups of the polyphenols in the inner skin might have been converted into $O\text{–}NH_4^+$ to incorporate ammonia gas into their molecular structures [19]. Additionally, the odor residual ratio for acetic acid gas incubated with the inner skin and compound paper was found to considerably decrease to 7.0 and 3.1% in 30 min, respectively. Therefore, the inner skin and compound paper were found to have a high deodorization capacity against acidic odor. The Weber–Fechner law indicates that the sensory intensity of an odor is in proportion to the logarithm of the odorant concentration. Therefore, when the concentration of an odor increases by 10-fold, the intensity of the odor can be perceived as two-fold. Conversely, the odor of one-tenth of the concentration may be perceived as 50% as intense as that of the original concentration. In conclusion, the ground inner skin and the compound paper produced using the skin were found to decrease the sensory intensity of the basic and acidic odors by approximately 50%. The inner skin of the chestnut showed a unique fibrous shape (Figure 5). Similarly, charcoal shows a high porosity, which is associated with its deodorizing effect [38]. The porosity of the inner skin could also be associated with its deodorizing effect. In contrast, the ground inner skin and compound paper showed negligible deodorizing activity toward acetaldehyde, a neutral odor. Further studies should be performed to determine the differences in the deodorizing effects of the inner chestnut skin and compound paper on different odor components.

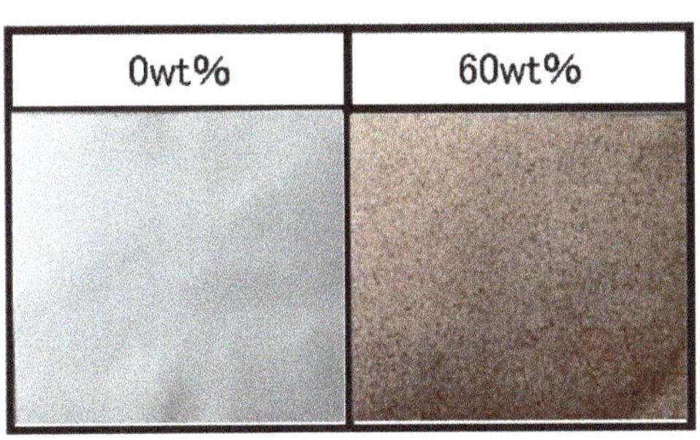

Figure 3. Compound paper (25 × 25 cm) comprising the inner skin of 'Porotan' chestnuts (60% weight).

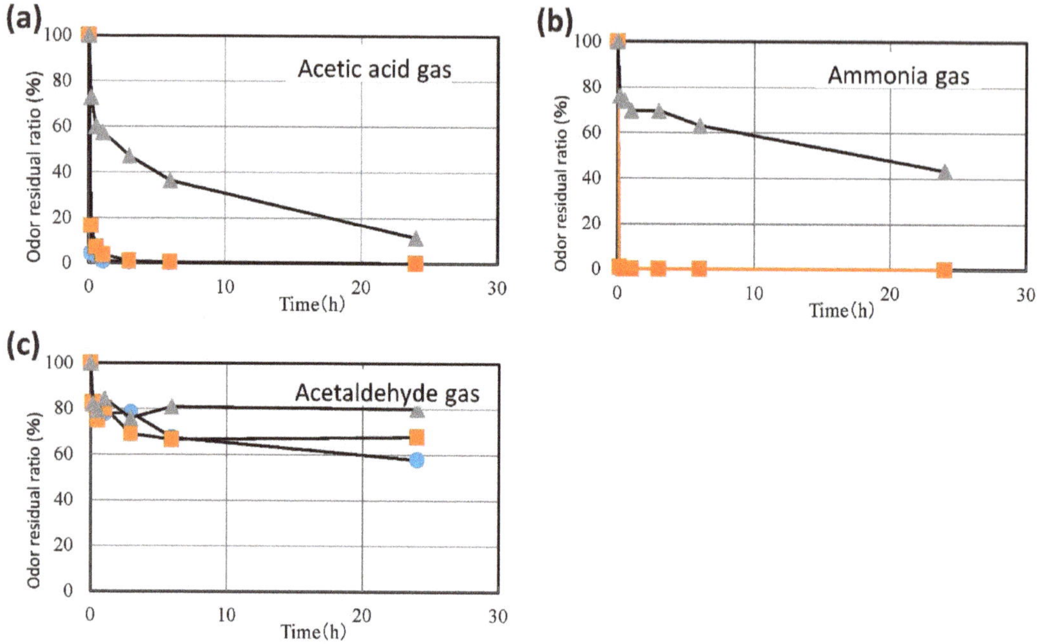

Figure 4. Deodorizing effects of the inner skin of 'Porotan' chestnuts and compound papers against various odorous gases. (**a**) Acetic acid gas, (**b**) ammonia gas, (**c**) acetaldehyde gas. The compounded papers were prepared by blending the ground inner skin of chestnuts (60% weight) with pulp fibers. •, compound papers; ■, inner skin; ▲, no sample against gas (control).

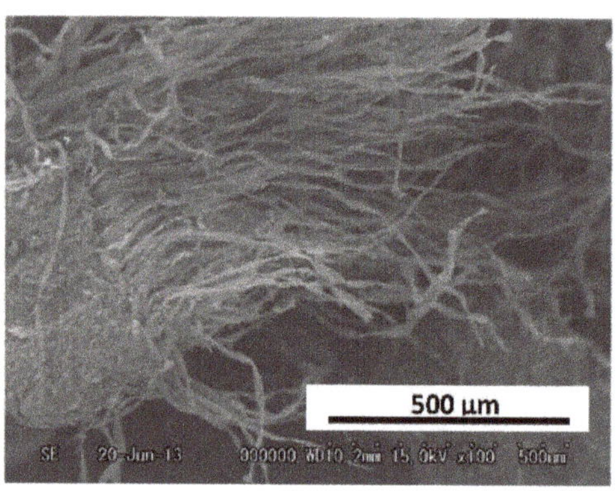

Figure 5. Scanning electron micrographs of skin of 'Porotan' (×500). Acceleration voltage: 20 kV.

4. Conclusions

To determine the potential re-utilization of the by-products of processed chestnut, chestnuts were separated into green and brown burs, shells, inner skin, kernel, and leaves; each part was powdered for analysis of its bioactive properties. Compound paper comprising inner skin was prepared, and its deodorizing properties were investigated. The analysis of the TPC, DPPH, and H-ORAC values showed that the inner skin possessed high antioxidant activity. Ground inner skin, green and brown burs, shells, and leaves showed excellent antibacterial activity against *S. aureus*, and brown burs showed antibacterial activity against *E. coli*. The inner skin considerably inhibited the growth of Gram-positive *S. aureus* and was found to possess high deodorant effect against basic and acidic odors. It also decreased the odor residual ratio of ammonia gas to 0% in only 10 min. A compound paper was prepared, comprising the inner skin of chestnuts, and it was found to possess a high deodorant effect. These results indicate that the inedible part of chestnuts has a very high potential for use in food and industrial products, and it is expected to be used in the future.

Author Contributions: Conceptualization, Y.T.; investigation, Y.T. and T.T.; data curation, Y.T. and T.T.; writing—original draft preparation, Y.T. and T.T.; writing—review and editing, Y.T.; visualization, Y.T.; supervision, Y.T.; project administration, Y.T.; funding acquisition, Y.T. Both authors have read and agreed to the published version of the manuscript.

Funding: This study was supported by the project "A Scheme to Revitalize Agriculture and Fisheries in Disaster Area through Deploying Highly Advanced Technology" in the Ministry of Agriculture, Forestry and Fisheries, and "KAKENHI, Grant-in-Aid for Scientific Research (B) (21H00808)", Japan.

Acknowledgments: We would like to thank Kent Yasuki and Tomoya Sato for performing experiments. We appreciate the Institute of Fruit Tree Science for the provision of samples of Porotan.

Conflicts of Interest: The authors declare no conflict of interest.

References

1. FAO. FAOSTAT. Available online: http://www.fao.org/faostat/en/?fbclid=IwAR0HODyYs-D3AYvKVIjfi3cUdlDYfkQjpT3tjtN5IE9KukB38Im_TUUxZMY#data/QC/visualize (accessed on 1 May 2021).
2. Tsurunaga, Y.; Saito, M.; Yasugi, K.; Takahashi, T. Technical Development of Value-Added Processed Products Using Frozen-stored'Porotan, a New Variety of Chestnut. *J. Jpn. Soc. Food Sci.* **2018**, *65*, 111–117. [CrossRef]
3. Vázquez, G.; Fontenla, E.; Santos, J.; Freire, M.; González-Álvarez, J.; Antorrena, G. Antioxidant activity and phenolic content of chestnut (*Castanea sativa*) shell and eucalyptus (*Eucalyptus globulus*) bark extracts. *Ind. Crop. Prod.* **2008**, *28*, 279–285. [CrossRef]
4. Hwang, J.-Y.; Hwang, I.-K.; Park, J.-B. Analysis of physicochemical factors related to the automatic pellicle removal in Korean chestnut (*Castanea crenata*). *J. Agric. Food Chem.* **2001**, *49*, 6045–6049. [CrossRef]
5. An, J.-Y.; Wang, L.-T.; Lv, M.-J.; Wang, J.-D.; Cai, Z.-H.; Wang, Y.-Q.; Zhang, S.; Yang, Q.; Fu, Y.-J. An efficiency strategy for extraction and recovery of ellagic acid from waste chestnut shell and its biological activity evaluation. *Microchem. J.* **2021**, *160*, 105616. [CrossRef]
6. Youn, U.-Y.; Shon, M.-S.; Kim, G.-N.; Katagiri, R.; Harata, K.; Ishida, Y.; Lee, S.-C. Antioxidant and anti-adipogenic activities of chestnut (*Castanea crenata*) byproducts. *Food Sci. Biotechnol.* **2016**, *25*, 1169–1174. [CrossRef] [PubMed]
7. Kawasaki, K.; Shiromizu, T.; Katsui, M. Suppressive effect of polyphenols extracted from chestnut skins on lipid absorption in rats. *Nippon Shokuhin Kagaku Kogaku Kaishi = J. Jpn. Soc. Food Sci. Technol.* **2009**, *56*, 545–548. [CrossRef]
8. Haslam, E. Natural polyphenols (vegetable tannins) as drugs: Possible modes of action. *J. Nat. Prod.* **1996**, *59*, 205–215. [CrossRef] [PubMed]
9. Kumari, M.; Jain, S. Tannins: An antinutrient with positive effect to manage diabetes. *Res. J. Recent Sci. ISSN* **2012**, *2277*, 2502.
10. Gupta, P.; Mohammad, T.; Khan, P.; Alajmi, M.F.; Hussain, A.; Rehman, M.T.; Hassan, M.I. Evaluation of ellagic acid as an inhibitor of sphingosine kinase 1: A targeted approach towards anticancer therapy. *Biomed. Pharmacother.* **2019**, *118*, 109245. [CrossRef]
11. Lee, N.-K.; Jung, B.S.; Yu, H.H.; Kim, J.-S.; Paik, H.-D. The impact of antimicrobial effect of chestnut inner shell extracts against Campylobacter jejuni in chicken meat. *LWT-Food Sci. Technol.* **2016**, *65*, 746–750. [CrossRef]
12. Misuri, F.; Marri, L. Antibacterial activity of wood distillate from residual virgin chestnut biomass. *Eur. J. Wood Wood Prod.* **2021**, *79*, 237–239. [CrossRef]
13. Ham, J.-S.; Kim, H.-Y.; Lim, S.-T. Antioxidant and deodorizing activities of phenolic components in chestnut inner shell extracts. *Ind. Crop. Prod.* **2015**, *73*, 99–105. [CrossRef]
14. Belwal, T.; Huang, H.; Li, L.; Duan, Z.; Zhang, X.; Aalim, H.; Luo, Z. Optimization model for ultrasonic-assisted and scale-up extraction of anthocyanins from *Pyrus communis* 'Starkrimson' fruit peel. *Food Chem.* **2019**, *297*, 124993. [CrossRef] [PubMed]
15. Takada, N.; Nishio, S.; Yamada, M.; Sawamura, Y.; Sato, A.; Hirabayashi, T.; Saito, T. Inheritance of the easy-peeling pellicle trait of Japanese chestnut cultivar Porotan. *HortScience* **2012**, *47*, 845–847. [CrossRef]
16. Saito, T.; Kotobuki, K.; Sawamura, Y.; Abe, K.; Terai, O.; Shoda, M.; Takada, N.; Sato, Y.; Hirabayashi, T.; Sato, A. *New Japanese Chestnut Cultivar 'Porotan'*; Bulletin of the National Institute of Fruit Tree Science: Tsukuba, Japan, 2009; pp. 1–9.
17. Takahashi, T.; Aso, Y.; Kasai, W.; Kondo, T. Effect of light irradiation on the antibacterial activity of compounded papers containing wasted tea leaves. *J. Wood Sci.* **2010**, *56*, 299–306. [CrossRef]

18. Takahashi, T.; Aso, Y.; Kasai, W.; Kondo, T. Improving the antibacterial activity against *Staphylococcus aureus* of composite sheets containing wasted tea leaves by roasting. *J. Wood Sci.* **2010**, *56*, 403–410. [CrossRef]
19. Takahashi, T.; Aso, Y.; Kasai, W.; Kondo, T. Synergetic deodorant effect and antibacterial activity of composite paper containing waste tea leaves. *J. Wood Sci.* **2011**, *57*, 308–316. [CrossRef]
20. Takahashi, T.; Tsurunaga, Y.; Schmidt, W.F.; Yoshino, K. Functional evaluation of horse chestnut seed and its application in the production of compounded paper for effective utilization of an untapped resource. *J. Wood Sci.* **2017**, *63*, 484–495. [CrossRef]
21. Swain, T.; Hillis, W. The phenolic constituents of *Prunus domestica*. I.—The quantitative analysis of phenolic constituents. *J. Sci Food Agric.* **1959**, *10*, 63–68. [CrossRef]
22. Katsube, T.; Tsurunaga, Y.; Sugiyama, M.; Furuno, T.; Yamasaki, Y. Effect of air-drying temperature on antioxidant capacity and stability of polyphenolic compounds in mulberry (*Morus alba* L.) leaves. *Food Chem.* **2009**, *113*, 964–969. [CrossRef]
23. Watanabe, J.; Oki, T.; Takebayashi, J.; Yamasaki, K.; Takano-Ishikawa, Y.; Hino, A.; Yasui, A. Method validation by interlaboratory studies of improved hydrophilic oxygen radical absorbance capacity methods for the determination of antioxidant capacities of antioxidant solutions and food extracts. *Anal. Sci.* **2012**, *28*, 159–165. [CrossRef]
24. Takahashi, T.; Aso, Y.; Yoshino, K. Incorporation of photocatalytic function into nonwoven polyester fabrics via impregnation with peroxo titanic acid solution. *J. Mater. Sci.* **2013**, *48*, 8199–8208. [CrossRef]
25. Takahashi, T.; Oowaki, M.; Onohara, Y.; Yanagi, Y. Deodorant performance of titanium dioxide-added acrylic/cellulose diacetate blended fibers. *Text. Res. J.* **2013**, *83*, 800–812. [CrossRef]
26. Tuyen, P.T.; Xuan, T.D.; Khang, D.T.; Ahmad, A.; Quan, N.V.; Tu Anh, T.T.; Anh, L.H.; Minh, T.N. Phenolic compositions and antioxidant properties in bark, flower, inner skin, kernel and leaf extracts of *Castanea crenata* Sieb. et Zucc. *Antioxidants* **2017**, *6*, 31 [CrossRef]
27. Schaich, K.M.; Tian, X.; Xie, J. Hurdles and pitfalls in measuring antioxidant efficacy: A critical evaluation of ABTS, DPPH, and ORAC assays. *J. Funct. Foods* **2015**, *14*, 111–125. [CrossRef]
28. Vekiari, S.A.; Gordon, M.H.; Garcia-Macias, P.; Labrinea, H. Extraction and determination of ellagic acid contentin chestnut bark and fruit. *Food Chem.* **2008**, *110*, 1007–1011. [CrossRef] [PubMed]
29. Li, D.; Zhang, X.; Qu, H.; Li, L.; Mao, B.; Xu, Y.; Lin, X.; Luo, Z. Delaying the biosynthesis of aromatic secondary metabolites in postharvest strawberry fruit exposed to elevated CO_2 atmosphere. *Food Chem.* **2020**, *306*, 125611. [CrossRef]
30. Vanella, L.; Di Giacomo, C.; Acquaviva, R.; Barbagallo, I.; Li Volti, G.; Cardile, V.; Abraham, N.G.; Sorrenti, V. Effects of ellagic acid on angiogenic factors in prostate cancer cells. *Cancers* **2013**, *5*, 726–738. [CrossRef]
31. Nankar, R.P.; Doble, M. Hybrid drug combination: Anti-diabetic treatment of type 2 diabetic Wistar rats with combination of ellagic acid and pioglitazone. *Phytomedicine* **2017**, *37*, 4–9. [CrossRef]
32. Liu, Y.; Liu, Y.; Liu, Y.; Liu, H.; Shang, Y. Evaluating effects of ellagic acid on the quality of kumquat fruits during storage. *Sci. Hortic.* **2018**, *227*, 244–254. [CrossRef]
33. Noormandi, A.; Dabaghzadeh, F. Effects of green tea on Escherichia coli as a uropathogen. *J. Tradit. Complement. Med.* **2015**, *5*, 15–20. [CrossRef] [PubMed]
34. Lee, Y.-L.; Cesario, T.; Wang, Y.; Shanbrom, E.; Thrupp, L. Antibacterial activity of vegetables and juices. *Nutrition* **2003**, *19*, 994–996. [CrossRef] [PubMed]
35. Ikigai, H.; Nakae, T.; Hara, Y.; Shimamura, T. Bactericidal catechins damage the lipid bilayer. *Biochim. Biophys. Acta (BBA) Biomembr.* **1993**, *1147*, 132–136. [CrossRef]
36. Cho, Y.; Schiller, N.; Kahng, H.; Oh, K. Cellular responses and proteomic analysis of Escherichia coli exposed to green tea polyphenols. *Curr. Microbiol.* **2007**, *55*, 501–506. [CrossRef] [PubMed]
37. JIS-L-1902. *Determination of Antibacterial Activity and Efficacy on Textile Products*; Japanese Standard Association: Tokyo, Japan, 2015; pp. 1–37. (In Japanese)
38. Eshtiaghi, M.N.; Kuldiloke, J.; Yoswathana, N. Deodorization of coconut oil using activated charcoal and charcoal regeneration. *J. Food Agric. Environ.* **2012**, *10*, 178–181.

foods MDPI

rticle

Enzymatic Digestion of Calf Fleshing Meat By-Products: Antioxidant and Anti-Tyrosinase Activity of Protein Hydrolysates, and Identification of Fatty Acids

ullia Tedeschi [1,*], Cecilia Anzani [1], Maura Ferri [2,3], Silvia Marzocchi [4], Maria Fiorenza Caboni [4], Stefania Monari [2] nd Annalisa Tassoni [2,*]

1 Department of Food and Drug, University of Parma, Parco Area Delle Scienze 49/a, 43124 Parma, Italy; anzani.cecilia@gmail.com
2 Department of Biological, Geological, Environmental Science, University of Bologna, Via Irnerio 42, 40126 Bologna, Italy; maura.ferri@unibo.it (M.F.); stefania.monari2@unibo.it (S.M.)
3 Department of Civil, Chemical, Environmental and Materials Engineering, University of Bologna, Via Terracini 28, 40131 Bologna, Italy
4 Department of Agricultural and Food Sciences and Technologies, University of Bologna, Piazza Goidanich 60, 47521 Cesena, Italy; silvia.marzocchi4@unibo.it (S.M.); maria.caboni@unibo.it (M.F.C.)
* Correspondence: tullia.tedeschi@unipr.it (T.T.); annalisa.tassoni2@unibo.it (A.T.); Tel: +39-0521-905406 (T.T.); +39-051-2091278 (A.T.)

Abstract: The food waste reduction through an efficient recovery of its valuable building molecules has become an important topic with a positive effect on the economy and the environment. In this work, the revalorization of slaughterhouse calf fleshing meat through its enzymatic hydrolysis is proposed. The proteolytic activity of 11 enzymes was initially screened and the four most efficient enzymes (papain, trypsin, pancreatin, and bromelain) were selected. The molecular profiling of the different protein/peptide fractions by the Linear Trap Quadrupole-OrbiTrap technique showed compositional differences due to the specificity of the enzymes' cleavage sites. In order to find a potential reuse of these hydrolysates, the analysis of antioxidant and, for the first time on fleshing meat hydrolysates, of anti-tyrosinase activities, was performed. Papain-digested samples were those showing the highest inhibition activity of tyrosinase enzyme (55.6%) as well as the highest antioxidant activity (3.52 g TEAC/L). In addition, the composition analysis of the lipid fraction was performed. The mono-unsaturated fatty acids resulted to be the most abundant lipid in all the samples with the exception of pancreatin-treated hydrolysates in which poly-unsaturated fatty acids were predominant. The present results seemed to support a possible valorization of isolated fractions from calf fleshing by-products, as food or feed ingredients, by the implementation of fraction isolation within the meat-processing pipeline.

Keywords: by-products valorization; protease; bioactive peptides; protein hydrolysates; fatty acids

1. Introduction

The recovery of food waste and by-products has become a hot topic, because of the rising global demand of animal protein-based food products up to 70% by 2050 [1]. This fact is due to the population growth and to the adoption, in underdeveloped countries, of Westernized eating habits. Furthermore, this expected increase of animal protein product consumption leads to a simultaneous increase of the production of food-related waste [2]. Interestingly, depending on animal species, after the slaughtering, only 30–40% of meat-processing products is used for human consumption [3]. Particularly in the case of beef, that 44% is considered as "meat" and the remaining 56%, usually discarded, is composed of non-edible (e.g., specified risked materials, SRM) and edible parts (e.g., offal) [4]. The incorrect utilization or the non-utilization of these by-products causes loss of potential revenues and rising disposal costs for the producing companies [5]. An efficient and sustainable utilization of these by-products will have a positive economic and environmental impact. Moreover, these by-products could be exploited as a cheap source of valuable compounds, which could be recovered and recycled as functional additives or ingredients in different food or feed products [6].

Beef fleshing meat by-product is represented by the meat portion, which is directly attached to the bovine hides. This part is usually discarded and not recovered despite its potential as a valuable source of proteins. Different strategies have been proposed [7] for the recovery of fleshing meat. Previous studies reported the use of the enzymatic hydrolysis on fleshing meat from different animal species [8–11] with or without the chemical treatments of tanning process. More recently, the activity of six proteolytic enzymes was compared, and the protein fraction composition characterised [12]. The meat by-product hydrolysates could find valuable industrial applications in the food sector as they constitute a notable source of nutrients like amino acids, minerals, and vitamins, they can be used to improve technological functions (flavouring compounds, water binding agents, emulsifiers), enhance sensory qualities (colour, texture, flavour), and increase the shelf life [13]. Moreover, a prevalent fraction of the meat hydrolysates is constituted by bioactive peptides, which could be used for the production of functional ingredients. Bioactive peptides are defined as a short sequence of amino acids, which are not active within the structure of the parent protein but show a positive effect on the system of the body once released [14]. During the proteolysis, a high number of peptides are liberated, but only a few of them can exhibit determined health benefits and can be defined as bioactive peptides [15]. Some examples of bioactivities, which can affect the health of the consumers, are anti-hypertensive, antioxidant, anti-inflammatory, antimicrobial, anticancer, anti-thrombic, and opioid-like [14]. Oxidation in food is directly linked to its deterioration and, in particular, it affects lipids, carbohydrates, and proteins. In meat products, the most important effects of deterioration are colour changes, microbial growth, and lipid oxidation [16]. To prevent the oxidation in food, synthetic antioxidants are usually added to retard lipid oxidation [17] but their utilization is under strict regulation as they might be associated with health risk. On the other hand, enzymatic browning due to tyrosinase activity may further result in discolouration and reducing the nutritional value of foods [18]. Therefore, nowadays, industries are searching for tyrosinase inhibitors to be used as browning preventing agents in nutraceutical and functional food sectors [19] or as anti-aging ingredients in cosmetic field [20]. This biological activity was usually ascribed to plant secondary metabolites [18,21,22] and, recently, also to proteins and peptides coming from both animals and plants [23–25]. Anti-tyrosinase capacity of fleshing meat hydrolysates is reported here for the first time and could prove to have good possibilities of future exploitation already in the meat processing industries. In fact, the protein hydrolysates coming from meat by-products and showing anti-browning capacity could be valuable, both as preservatives for meat-based food products, as well as additives for active food packaging.

As regards fatty acid total content, saturated fatty acids are the preponderant in bovine meat. Because of the hydrogenation process of unsaturated fatty acids that make them saturated that occurs in bovine rumen due to its microflora, polyunsaturated fatty acid content is very low in bovine meat [26]. A study conducted on calf, young bull, and cow meat confirmed this fatty acid development [27]. Animal species, breed, age, and farmed technique can affect fatty acids composition [28].

In view of the reduction of large amounts of by-products and wastes generated during meat processing, the aim of this study was to test the activity of eleven proteases in different reaction conditions for the hydrolysis of bovine fleshing meat and, after the selection of the four most efficient enzymes, to determine the molecular profile of the protein and lipid fractions of obtained hydrolysates. The peptide hydrolysates were also tested in vitro to assess their antioxidant and, for the first time, their anti-tyrosinase activities in view of their possible exploitation in the food and feed industrial sectors.

2. Materials and Methods

2.1. Reagents and Solvents

Deionized water was obtained from a Millipore Alpha Q-Waters purification system (Billerica, MA, USA). Sodium dihydrogen phosphate, hydrochloric acid, acetonitrile, DL-cystine, dichloromethane, N-acetyl-L-cysteine, sodium tetraborate decahydrate, DL-isoleucine and DL-norleucine, 30% hydrogen peroxide solution, sulfuric acid, formic acid, trifluoroacetic acid, 2′-azino-bis-3-ethylbenzothiazoline-6-sulfonic acid (ABTS), potassium persulfate, Trolox, tyrosinase from mushroom, and L-DOPA were purchased from Sigma-Aldrich (St. Louis, MO, USA). Copper oxide was obtained from Carlo Erba (Milan, Italy). O-phthalaldehyde and boric acid were bought from Fluka (Buchs, Switzerland). Sodium dodecyl sulfate was purchased from Biorad (Hercules, CA, USA). Kjeldahl tablets defoamers and catalyst 3.5 g/tablet were purchased from Merck (Darmstadt, Germany). AccQ-Fluor reagent kit was obtained from Waters (Milford, MA, USA).

2.2. Raw Material

Samples of calf fleshing were provided by Inalca Industria Alimentare Carni SpA (Castelvetro di Modena, Italy). The samples (three biological replicates of about 600 gFW each collected on a single slaughter day) were obtained from commercial breeds slaughtered in an abattoir at Inalca and transported to the University laboratories on the day of the slaughtering. The samples were grounded and homogenized in a kitchen blender (Kenwood FP735 1000 W, Tokyo, Japan) at 3 °C. The samples were divided in 5.0 ± 0.3 gFW (grams of fresh weight) aliquots and immediately frozen at −80 °C until further analyses.

2.3. Raw Material Chemical Composition

Moisture content of fleshing was determined using the official AOAC (Association of Official Analytical Chemists) method [29] according to Anzani et al. [12].

2.4. Proteases Digestion

Just before hydrolysis, 10 mL of 10 mM sodium-phosphate buffer (pH 7.5) were added to 5 gFW of fleshing; the final pH of the digestion mixture was 7.2 ± 0.2. Initially, the proteolytic activity of 11 different commercial proteases were initially tested following these conditions: 2% E/S ratio (g enzyme/gFW sample), 4 h of incubation time, pH 7.2–7.4 (with the exception of pepsin, which was tested at pH 3.0, in water acidified by chloridric acid), 140 rpm stirring, and specific temperature enzyme according to manufacturer's instructions: Alcalase 1.5 MG (60 °C, Novozymes A/S, Denmark), protamex (60 °C, Novozymes A/S, Denmark), papain from *Carica papaya* (60 °C, Sigma-Aldrich, Milan, Italy), neutrase 1.5 MG (50 °C, Novozymes A/S, Denmark), flavourzyme 500 MG (50 °C, Novozymes A/S, Denmark), pancreatin from porcine pancreas (37 °C, Sigma-Aldrich, Milan, Italy), trypsin (type A, 13,000–20,000 U/mg, 37 °C, Sigma-Aldrich, Milan, Italy), trypsin from porcine pancreas (type B, 1000–2000 U/mg, 37 °C, Sigma-Aldrich, Milan, Italy), bromelain from pineapple stem (37 °C, Sigma-Aldrich, Milan, Italy), α-chymotrypsin from bovine pancreas (37 °C, Sigma-Aldrich, Milan, Italy), and pepsin from porcine gastric mucosa (37 °C, pH 3.0, Sigma-Aldrich, Milan, Italy). Control samples without any enzyme addition (ND, not digested) were performed by incubating the samples for 4 h, 140 rpm stirring, at each incubation temperature (ND 60 °C, 50 °C and 37 °C). The enzyme was inactivated by boiling the samples in a water bath for 5 min. The samples were cooled on ice and centrifuged at 5000 rpm for 15 min at 4 °C (Eppendorf centrifuge 5804 R, rotor A 4-44, Hamburg, Germany). After the centrifugation, three different fractions were observed: Pellet (residual solid material), intermediate liquid fraction (containing digested proteins/peptides/amino acids), upper fraction (containing lipids). The three fractions were separately collected and stored at −20 °C until following analyses.

Because of the gelatine formation in ND (60 °C, 4 h), α-chymotrypsin (60 °C, 4 h), and pepsin (37 °C, 4 h) samples, it was not possible to separate the three sub-fractions in these treatments, therefore they were discarded.

To optimize the E/S ratio and the time of incubation, several digestion treatments were performed for each type of enzyme. The four best digestion conditions were selected, yielding the highest amount of hydrolyzed proteins in the intermediate fraction measured as g of bovine serum albumin equivalents per liter (g BSA eq/L) [30]. The complete list of tested hydrolysis conditions and related yields is summarized in Supplementary Table S1.

2.5. Quantification of Free and Total Amino Acids, SDS-PAGE and Degree of Hydrolysis (DH) Analyses

The free amino acid determination in intermediate protein-based fractions, performed by high-performance liquid chromatography with fluorescence detection (HPLC/FLD) after AccQ•Tag derivatization, the SDS-PAGE (Sodium Dodecyl Sulphate - PolyAcrylamide Gel Electrophoresis) analysis, and the DH analysis by (o-phthaldialdehyde) OPA method, were performed as reported by Anzani et al. [12]. The DH was calculated as follows:

$$DH\ (\%) = (N_{free}/N_{total}) \times 100 \quad (1)$$

where N_{free} represent the moles of free nitrogen atoms from α- amino groups of amino acids after hydrolysis measured by the OPA assay, and N_{total} represent the total moles of nitrogen atoms in solution before hydrolysis calculated by the ratio of total grams of proteins and the average residual amino acid molecular mass (MW 110 Da).

2.6. Peptide Identification by Linear Trap Quadrupole (LTQ)-Orbitrap Analyses

Peptide identification was performed by LTQ-Orbitrap analyses according to a method previously reported [12].

2.7. Fatty Acid Analysis

The fatty acid composition of upper fractions was determined as fatty acid methyl esters (FAMEs) by capillary gas chromatography analysis after alkaline treatment [31]. Methyl tridecanoate (C13:0, 2 mg/mL) was used as internal standard, and a GC 2010 Plus gas chromatograph (Shimadzu Corporation, Kyoto, Japan) equipped with a flame ionization detector (FID) and an AOC-20s auto sampler (Shimadzu Corporation) was used, according to Verardo et al. [32] with slight modifications. A BPX70 fused silica capillary column (10 m × 0.1 mm i.d. 0.2 μm film thickness; SGE Analytical Science, Ringwood, VIC. Australia) was used. The injector and flame ionization detector temperatures were set at 250 °C. Hydrogen was used as carrier gas at a flow rate of 0.8 mL/min. The oven temperature was held at 50 °C for 0.2 min, increased to 175 °C at 120 °C/min, held at 175 °C for 2 min, and finally increased from 175 to 220 °C at 20 °C/min. Samples were injected in split mode (0.3 μL) with a split ratio set at 1:100. Peak identification was accomplished by comparing peak retention time with GLC-463 standard mixture from Nu-Check (Elysian, MN, USA) and FAME 189-19 standard mixtures from Sigma-Aldrich Chemicals (St. Louis, MO, USA) and expressed as weight percentage of total FAMEs. FAMEs composition was measured in 2 replicates for each lipid extract (n = 6) and each analysis lasted 7 min.

2.8. Antioxidant and Anti-Tyrosinase Activity Evaluation

In all experiments, in vitro antioxidant activity of intermediate fractions was assessed by the 2′-azino-bis-3-ethylbenzothiazoline-6-sulfonic acid (ABTS) assay as reported by Ferri et al. [33]. The results were expressed as Trolox Equivalents Antioxidant Capacity (TEAC) by means of a dose–response calibration curve (between 0 and 2.5 μg of Trolox). Anti-tyrosinase capacity was measured by an optimized enzyme inhibition assay [24]. The brown color formation kinetic was detected (absorbance at 490 nm) in the presence of 10 U of tyrosinase, 2 mM L-DOPA (3,4-dihydroxy-L-phenylalanine), and the sample. The results were expressed as % of inhibition.

2.9. Statistical Analysis

All the treatments were performed in two biological replicates and two technical replicates were analyzed for each repetition. Statistically significant differences were analyzed among the same type of analytical data, related to different hydrolysate samples, by using one-way ANOVA followed by post hoc Tukey's multiple pairwise comparison ($p < 0.05$) (R Software, version 1.3.5, R Core Tram, Vienna, Austria).

3. Results and Discussion

3.1. Calf Fleshing Bulk Characterization

Non-treated calf fleshing meat samples (ND) were analyzed. Proteins, fat, and moisture percentages found were 16% (± 4), 14% (± 2), and 72% (± 3), respectively, over the total wet weight. Previous data obtained on the same initial by-product [12] showed the presence of a lower fat content. This difference could be ascribable to calf samples' biological variability, which could show a slightly different composition of the residual fleshing. Conversely, the distribution of total amino acids (Table 1) was similar to that reported by Anzani et al. [12], with Hyp, Glu, and Pro being the most abundant. In agreement with Rai et al. [34] and Bhaskar et al. [35], the presence of these amino acids suggested that fleshing is constituted by a great amount of collagen because of its proximity to the skin. A further important point of discussion is to evaluate the impact of the fleshing from a nutritional point of view. Therefore, the amount of essential amino acids was compared with that of a reference standard amino acid profile from raw lean beef [36]. Fleshing showed a lower amount of essential amino acids compared to the reference, demonstrating the lower nutritional value of fleshing compared to lean meat, given the presence of the collagen as fleshing main component. Collagen, as previously mentioned, is rich in Hyp, Pro, and Glu, and consequently caused a decrease of the essential amino acids in fleshing.

Table 1. High-performance liquid chromatography with fluorescence detection (HPLC-FLD) analysis of amino acid distribution in initial calf fleshing by-products given as percentage of the total amino acid pool.

Amino Acids	% on Total Amino Acids	Reference (%)
Hyp	12.1 ± 0.6	1.4
Asp	7.0 ± 0.5	9.4
Ser	3.2 ± 0.1	4.2
Glu	13.2 ± 0.7	15.6
Gly	8.2 ± 0.6	7.1
His	**2.2 ± 0.3**	**3.4**
Arg	6.3 ± 0.5	6.8
Thr	**3.4 ± 0.3**	**4.0**
Ala	5.3 ± 0.8	6.5
Pro	10.7 ± 1.2	5.3
Tyr	2.7 ± 0.4	3.2
Val	**3.3 ± 0.2**	**5.1**
Met	**3.4 ± 0.3**	**2.7**
Lys	**6.6 ± 0.4**	**8.6**
Ile	**3.3 ± 0.3**	**4.6**
Leu	**5.8 ± 0.5**	**8.1**
Phe	**3.4 ± 0.3**	**4.1**

Data in bold show the amounts of essential amino acids. The reference is represented by standard amino acid profile from raw lean beef [36]. Data (means ± SD) are expressed in (%).

3.2. Optimization of the Enzymatic Reaction

In preliminary digestion trials, 11 different proteases were tested on calf fleshing samples. To assess the efficiency of the digestion reactions, the hydrolyzed protein amount as well as the antioxidant activity were measured, by spectrophotometrical assays, in the intermediate liquid protein-based fraction of hydrolysates (Supplementary Table S1). After preliminary trials, the four best digestion conditions, yielding the highest amount of digested proteins, were selected and were, respectively: papain: 10% E/S, 2 h 60 °C; pancreatin 10% E/S, 2 h 37 °C; bromelain 10% E/S, 2 h 37 °C; trypsin (type B): 10% E/S, 4 h 37 °C. Overall, the highest amount of hydrolyzed proteins was equally measured in samples obtained with 10% papain and 10% trypsin B (77.31 and 77.65 g BSA eq/L, respectively) followed by bromelain (68.88 g BSA eq/L) and pancreatin (58.76 g BSA eq/L) (Figure 1a, Supplementary Table S1).

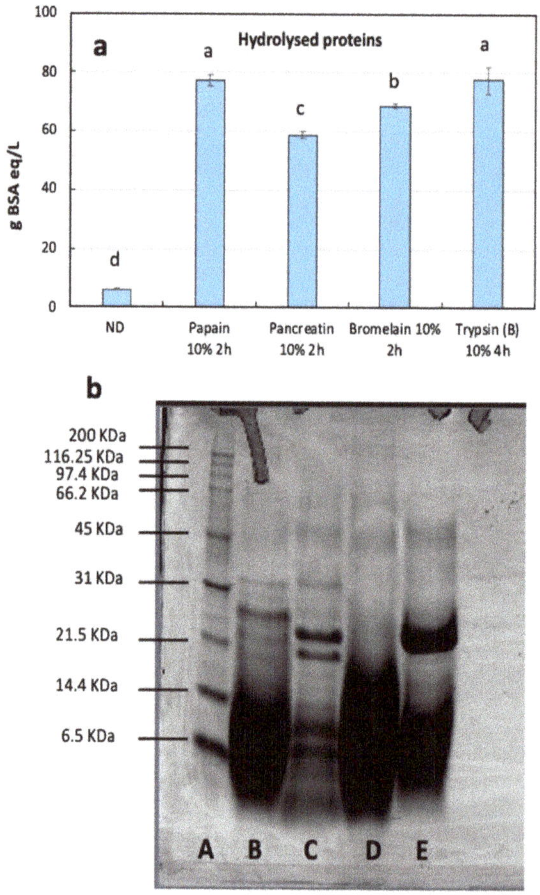

Figure 1. Optimized hydrolysis. (**a**) Quantification of hydrolyzed proteins (g BSA eq/L) in intermediate fractions obtained after calf fleshing hydrolysis at different digestion conditions. ND, average of not-digested control at 50 °C and 37 °C. Statistically significant differences among different hydrolysate samples were indicated by different letters, according to one-way ANOVA followed by post hoc Tukey's multiple pairwise comparison ($p < 0.05$). Data are the mean ± SD ($n = 4$). (**b**) SDS-PAGE (Sodium Dodecyl Sulphate - PolyAcrylamide Gel Electrophoresis) profile of samples: A, standard marker; B, papain hydrolysate; C, pancreatin hydrolysate; D, bromelain hydrolysate; E, trypsin (B) hydrolysate.

The SDS-PAGE sample profile (Figure 1b) showed that the hydrolysates were constituted by peptides with a molecular weight below 31 kDa. However, there were some differences among the samples as consequence of the different cleavage sites of each enzyme, which could attack the sequence within specific amino acid residues or be unspecific. Of the selected enzymes, bromelain does not have a specific cleavage site while papain is a cysteine endopeptidase, hydrolyzing proteins, peptides, amides, and esters of amino acids and peptides, with the specific target of Arg, Lys, Glu, His, Gly, and Tyr bonds [37]. However, if the incubation is prolonged, further bonds are split. According to Olsen et al. [38], the cleavage site of trypsin is very specific cutting exclusively the C-terminal Arg or Lys. On the contrary, pancreatin cleavage site is not specific being a mixture of different type of enzymes such as protease, amylase, and lipase. However, the profile of the sample treated with pancreatin and trypsin type B resulted almost similar, with a blurred band lower than 14.4 kDa and 2 bands around 21.5 kDa (probably because of the high concentration in the sample treated with trypsin showed only one blurred band) (Figure 1b, lanes C and E).

3.3. Protein Fraction Analysis

The degree of hydrolysis (DH), which is directly linked to the enzyme proteolytic activity, was determined on selected samples on the basis of the number of the cleaved peptide bonds and on the amount of free amino acids present in solution [39]. DH was determined by monitoring the reaction of the free amino groups with OPA depending on the protein material released in solution. The sample treated with papain showed the highest DH (70% ± 2), followed by bromelain and pancreatin (55% ± 1 and 60% ± 5, respectively), which resulted to be not significantly different ($p < 0.05$). The sample obtained with trypsin B showed the lowest DH (25% ± 0.4) possibly due to the highest specificity of this enzyme, which is able to cleave the sequence after Arg and Lys and therefore release long peptides and less free amino acids. The present DH results were higher than literature data; for example, Zhang et al. [9] treated bovine collagen with six different proteases and diverse pre-treatments, obtaining DH always lower than 20%. The content and distribution of free amino acids were also measured and are reported in Table 2.

Table 2. Free amino acid distribution of the hydrolysates obtained by enzymatic hydrolysis of calf fleshing meat.

Amino Acids	Papain	Pancreatin	Bromelain	Trypsin (B)
Hyp	3.46 ± 0.33 b	3.54 ± 0.07 b	2.79 ± 0.60 b	5.75 ± 0.30 a
Asp	0.14 ± 0.03 c	1.81 ± 0.01 a	0.81 ± 0.01 b	1.78 ± 0.36 a
Ser	3.55 ± 0.06 a	4.17 ± 0.66 a	3.73 ± 0.20 a	1.75 ± 0.01 b
Glu	4.66 ± 0.04 b	6.86 ± 0.25 a	2.44 ± 0.37 c	2.77 ± 0.05 c
Gly	9.00 ± 0.02 a	1.67 ± 0.15 c	3.27 ± 0.13 b	1.57 ± 0.04 c
His	2.62 ± 0.06 c	5.80 ± 0.08 a	4.65 ± 0.16 b	2.47 ± 0.02 c
Arg	14.58 ± 0.29 a	16.63 ± 1.69 a	3.74 ± 0.34 b	14.27 ± 1.04 a
Thr	2.93 ± 0.08 b	3.79 ± 0.07 b	8.37 ± 0.24 a	3.48 ± 0.05 b
Ala	3.44 ± 0.03 b	3.70 ± 0.67 a	2.29 ± 0.09 b	3.28 ± 0.80 b
Pro	7.67 ± 0.04 b	1.32 ± 0.23 c	11.89 ± 0.62 a	1.36 ± 0.21 c
Tyr	7.89 ± 0.21 a	8.93 ± 0.86 a	4.96 ± 0.47 b	7.52 ± 0.73 a
Val	1.85 ± 0.09 c	4.43 ± 0.15 a	3.08 ± 0.01 b	3.46 ± 0.69 b
Met	3.93 ± 0.13 b	4.20 ± 0.41 b	4.55 ± 0.22 b	5.69 ± 0.03 a
Lys	15.36 ± 0.13 b	14.89 ± 1.84 b	10.55 ± 0.06 c	22.51 ± 1.89 a
Ile	3.96 ± 0.12 b	4.05 ± 0.20 b	19.26 ± 0.23 a	4.47 ± 0.69 b
Leu	11.20 ± 0.47 a	7.82 ± 0.33 c	9.04 ± 0.16 b	9.45 ± 0.38 b
Phe	3.75 ± 0.33 c	6.40 ± 0.58 b	4.59 ± 0.09 c	8.40 ± 0.45 a

Data (means ± SD) are expressed in (%). Different lowercase letters on the same row show significantly different values for samples digested with different enzyme determined one-way ANOVA followed by post hoc Tukey's multiple pairwise comparison ($p < 0.05$).

Particularly in the case of trypsin B and pancreatin digested samples, the most abundant amino acids were Arg (14.27 and 16.63%, respectively) and Lys (22.51 and 14.89%, respectively), according to the specificity of the cleavage site. The sample treated with papain was rich of Arg (14.58%), Lys (15.36%), Leu (11.20%), and Gly (9.00%), while that obtained with bromelain showed high levels of Ile (19.26%), Pro (11.89%), Lys (10.55%), Leu (9.04%), and Thr (8.37%).

Moreover, the hydrolysates were also analyzed by high-resolution mass spectrometry to identify the most abundant peptides present in the origin proteins (Table 3). In the pancreatin and papain hydrolysates most of the detected peptides derived from collagen. On the other side, in trypsin B-derived digestates the most abundant peptides were originated from muscular protein, such as myosin.

Table 3. Identification by Linear Trap Quadrupole-OrbiTrap of the main peptides in digestates obtained after enzymatic hydrolysis of calf fleshing meat, and of the proteins of origin. Proteins are listed according to the highest number of peptides generated [12]. AA, amino acids.

Enzyme	Identified Peptides	Parent Protein	Average Peptide Length (N°AA)
Papain	78	Collagen alpha-1(I) chain	14
	71	Actin, alpha skeletal muscle	11
	56	Collagen alpha-2(I) chain	13
	54	Collagen alpha-1(II) chain	12
	30	Myosin-1	11
Pancreatin	70	Titin	12
	57	Collagen alpha-1(I) chain	11
	53	Collagen alpha-2(I) chain	10
	29	Collagen, type III, alpha 1	14
	24	IgG heavy chain	18
Bromelain	24	Collagen alpha-1(I)	9
	17	Hemoglobin subunit beta	23
	15	Hemoglobin subunit alpha	25
	15	Collagen alpha-2(I) chain	9
	13	Myosin light chain 1/3, skeletal muscle isoform	13
Trypsin (B)	59	Titin	11
	46	Myosin-2	11
	28	Myosin-1	10
	23	Collagen alpha-1(I) chain	13
	10	Myosin-7	12

3.4. *Antioxidant and Anti-Tyrosinase Activity*

The antioxidant activity was assayed in intermediate fraction of all the samples (Supplementary Table S1) and the highest values were obtained using 10% papain (on average 3.22 g TEAC/L) (Supplementary Table S1, Figure 2a). These results are comparable with literature data by Ryder et al. [40] who obtained hydrolysates, from bovine connective tissue treated with different fungal proteases, showing an average Trolox equivalence lower than 10 mmol/L (equals to 2.5 g TEAC/L), increased to 45 mmol/L (equals to 11 g TEAC/L) by a more complex proteolytic protease preparation.

In addition, the papain-sample (10% for 2 h) also showed the highest inhibition activity (55.6%) of tyrosinase enzyme, while the sample digested with bromelain (10% for 2 h) showed the lowest (23.6%) (Figure 2b). To the best of the authors' knowledge, this is the first report on anti-tyrosinase activities of fleshing meat hydrolysates. A comparison can be done with tyrosinase inhibitory activity of a fish meat (*Scomber japonicus*) hydrolysed by subcritical water treatment, which shows best activities between 48% and 65.5% of inhibition depending on process conditions [41]. These results were, therefore, very promising, as nowadays, the interest in anti-tyrosinase ingredients is increasing among several industries [19,20]. Fleshing meat hydrolysates exerting this type of bioactivity can be exploited as browning-preventing agents already in the meat-processing chain, for example as preservative ingredients in food preparations, or as additives for the production of food active packaging.

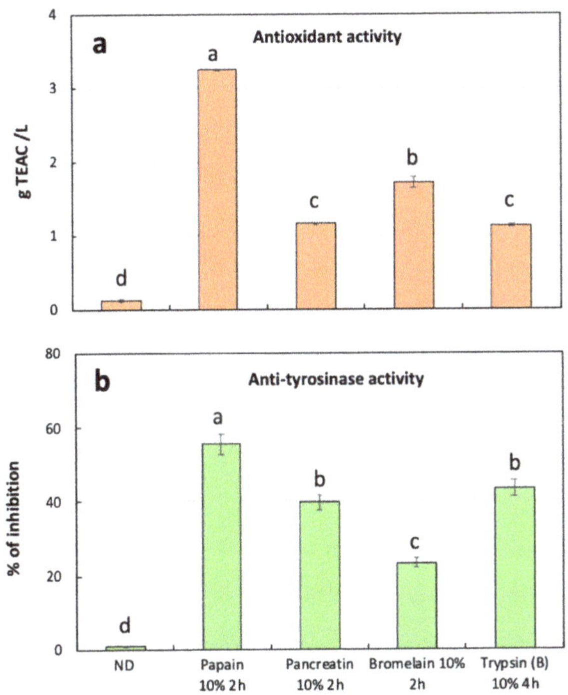

Figure 2. Quantification of antioxidant activity (**a**) and anti-tyrosinase activity (**b**) in intermediate fractions obtained after calf fleshing hydrolyses at different digestion conditions. ND, average of not digested control at 50 °C and 37 °C. Statistically significant differences among the same analysis were indicated by different letters, according to one-way ANOVA followed by post hoc Tukey's multiple pairwise comparison ($p < 0.05$). TEAC, Trolox equivalents antioxidant capacity. Data are the mean ± SD ($n = 4$).

3.5. Fatty Acid Composition

A total of 17 individual fatty acids were identified and quantified in the upper fraction of differently hydrolyzed samples. As shown in Table 4, oleic acid (C18:1 cis9) was the most abundant fatty acid in all digestates, ranging from about 39.59 to 42.80%. The second major fatty acid detected was palmitic acid (C16:0, 24–26%), followed by stearic acid (C18:0, 7–9%), palmitoleic acid (C16:1 cis, 5–6%), myristic acid (C14:0, 5–6%), and linoleic acid (C18:2 n6, 4–5%). The samples digested with four different enzymes differed for the fatty acid composition but not for the concentration of them, that did not show significant differences ($p < 0.05$).

As shown in Table 4, the sample digested with papain showed the most complete fatty acids composition, which was the same of sample digested with trypsin, except for C18:1 t9 that was absent. Samples treated with pancreatin and bromelain showed a more limited fatty acids composition. In particular, C12:0, C15:0, C16:1t, C17:0, C17:1, C18:1t9, C18:3n3, and C20:0 were absent in these two samples; C20:3n6 was present only in the sample digested with pancreatin. Saturated fatty acids (SFA) were present in considerable amounts, from 39.52 to 40.46%, whereas mono-unsaturated fatty acids (MUFA) were the class more abundant in all the samples ranging from 50.20 to 54.60%

without significant differences ($p < 0.05$) for both classes among the different samples. Polyunsaturated fatty acids (PUFA), instead, showed a significantly higher amount in the sample digested with pancreatin (10.03%) and in a range from 4.99 to 6.39% in the other three samples. The low PUFA content of the analyzed calf's by-products is related to animal nature, and rumen microflora in bovine rumen is responsible for the hydrogenation process, which takes place at unsaturated fatty acids, increasing the total SFA amount [26,27]. Compared to the literature [42–44], the results obtained in this work revealed lower amount of saturated and polyunsaturated fatty acids and higher concentration of monounsaturated fatty acids. In fact, in studies conducted by Baggio et al. [42], Muchenje et al. [43], and Brugiapaglia et al. [44], SFA concentrations were, respectively, 47–48%, 44–45% and 46–49%; MUFA concentrations were 39–40%, 34–36%, and 32–42%; and PUFA concentrations were 12–14%, 19–22%, and 41–22%.

Table 4. Fatty acids (FA) identification and quantification of the hydrolysates obtained by enzymatic hydrolysis of fleshing meat.

FA	Papain	Pancreatin	Bromelain	Trypsin (B)
C12:0	0.69 ± 0.00	n.d.	n.d	0.51 ± 0.02
C14:0	5.82 ± 0.09	5.88 ± 0.04	5.66 ± 0.30	5.41 ± 0.33
C14:1c	2.30 ± 0.19	2.67 ± 0.56	3.79 ± 0.15	2.14 ± 0.01
C15:0	0.31 ± 0.01	n.d	n.d	0.28 ± 0.00
C16:0	24.66 ± 0.03	23.99 ± 1.95	26.55 ± 0.38	24.72 ± 0.19
C16:1 t	0.37 ± 0.02	n.d	n.d	0.27 ± 0.07
C16:1 c	5.80 ± 0.02	5.89 ± 0.60	5.88 ± 0.96	5.92 ± 0.28
C17:0	0.58 ± 0.01	n.d	n.d	0.59 ± 0.07
C17:1	0,61 ± 0.04	n.d	n.d	0.64 ± 0.10
C18:0	7.77 ± 0.03	9.91 ± 0.66	8.21 ± 0.47	7.59 ± 0.40
C18:1 t9	1.56 ± 0.11	n.d	n.d	n.d
C18:1 c9	40.52 ± 0.02	39.59 ± 4.69	42.80 ± 1.97	41.61 ± 1.37
C18:1 c11	1.99 ± 0.14	2.06 ± 0.42	2.12 ± 0.32	2.13 ± 0.10
C18:2 n6	4.91 ± 0.15	4.78 ± 0.38	4.99 ± 0.95	4.38 ± 0.60
C18:3 n3	0.71 ± 0.08	n.d	n.d	0.90 ± 0.08
C20:0	0.64 ± 0.06 ^{a}	n.d	n.d	0.42 ± 0.01 ^{b}
C20:3 n6	0.76 ± 0.66	5.24 ± 0.77	n.d	0.85 ± 0.10
SFA	40.46 ± 0.08	39.77 ± 0.75	40.41 ± 1.16	39.52 ± 0.98
MUFA	53.16 ± 0.51	50.20 ± 3.15	54.60 ± 2.11	54.36 ± 1.40
PUFA	6.39 ± 0.59 ^{b}	10.03 ± 2.39 ^{a}	4.99 ± 0.95 ^{c}	6.12 ± 0.42 ^{b}

SFA, saturated fatty acid; MUFA, monounsaturated fatty acid; PUFA, polyunsaturated fatty acid. n.d., not determined. Data (means ± SD) are expressed in mg FA/100 mg fatty acid methyl esters (FAME) (%). Different letters on the same row show significantly different values for samples digested with different enzyme determined by one-way ANOVA followed by post hoc Tukey's multiple pairwise comparison ($p < 0.05$).

4. Conclusions

In the present paper, the best four enzyme protocols for the hydrolysis of calf fleshing meat were selected and optimized. These were: Papain (10% E/S, for 2 h at 60 °C); pancreatin (10% E/S, for 2 h at 37 °C); bromelain (10% E/S, for 2 h at 37 °C); and trypsin (type B) (10% E/S, for 4 h at 37 °C). SDS-PAGE profile, free-amino acid contents, DH, and peptide identification results pointed out that the differences between the hydrolysates were due to the specificity of the enzyme's cleavage site. In the case of the fatty acid composition, MUFA were the most abundant ranging from 50.20 to 54.60%; whereas the digestate obtained with pancreatin showed the highest amount of PUFA (10.03%). The obtained protein-based samples also showed both antioxidant and anti-tyrosinase biological activities with papain-digested samples displaying the highest bioactivities (up to 3.52 g TEAC/L antioxidant capacity and to 55.6% tyrosinase inhibition). These results demonstrated the potential use of the enzymatic hydrolysis as a way of recovering calf fleshing meat by the production of hydrolysates reach in proteins, peptides, and free amino acids with anti-oxidant and anti-tyrosinase activities, and of good quality fatty acids fractions like n-3 and n-6 series. As no solvents nor chemicals were used along the exploitation process, the present results suggest that a possible future valorization of isolated fractions from calf fleshing by-products, as food or feed ingredients, is feasible and could be implemented short term within the meat-processing pipeline.

Supplementary Materials: The following are available online at https://www.mdpi.com/article/10.3390/foods10040755/s1, Table S1: Preliminary optimisation of calf fleshing meat digestion conditions and selection of most efficient proteases.

Author Contributions: Conceptualization, A.T. and T.T.; methodology, A.T., S.M. (Silvia Marzocchi) and T.T.; formal analysis, C.A., A.T., M.F., S.M. (Silvia Marzocchi), S.M. (Stefania Monari) and T.T.; investigation, C.A., M.F., S.M. (Silvia Marzocchi), S.M. (Stefania Monari) and T.T.; data curation, C.A., M.F., S.M. (Silvia Marzocchi), S.M. (Stefania Monari), A.T. and T.T.; writing—original draft preparation, A.T., S.M. (Silvia Marzocchi) and T.T.; writing—review and editing, A.T., M.F., S.M. (Silvia Marzocchi) and T.T.; supervision, A.T., M.F.C. and T.T.; funding acquisition, A.T., M.F.C. and T.T. All authors have read and agreed to the published version of the manuscript.

Funding: This research was funded by Italian Ministry of Education, University and Research, National Project Cluster Agrifood SO.FI.A. (SOstenibilità della FIliera Agroalimentare) (2013–2017, Grant Number CTN01_00230_450760).

Institutional Review Board Statement: Not applicable.

Informed Consent Statement: Not applicable.

Data Availability Statement: All data are available within the manuscript and Supplmentary Material.

Acknowledgments: Authors wish to thank Inalca Industria Alimentare Carni SpA (Castelvetro di Modena, Italy) that provided support in the form of meat sample research materials.

Conflicts of Interest: The authors declare no conflict of interest.

References

1. FAO. The Future of Food and Agriculture: Alternative Pathways to 2050. Available online: http://tools.foodsecurityportal.org/fao-global-perspectives-studies2019 (accessed on 20 January 2021).
2. FAO. SAVE FOOD: Global Initiative on Food Loss and Waste Reduction. 2021. Available online: http://www.fao.org/save-food/resources/en/ (accessed on 20 January 2021).
3. Ockerman, H.W.; Hansen, C.L. Introduction and history of processing animal by-products. In *Animal By-Products Processing and Utilization*; Ockerman, H.W., Hansen, C.L., Eds.; CRC Press: Boca Raton, FL, USA, 2000; pp. 1–22.
4. Russ, W.; Meyer-Pittroff, R. Utilizing waste products from the food production and processing industries. *Crit. Rev. Food Sci. Nutr.* **2004**, *44*, 57–62. [CrossRef] [PubMed]
5. Jayathilakan, K.; Sultana, K.; Radhakrishna, K.; Bawa, A.S. Utilization of by-products and waste materials from meat, poultry and fish processing industries: a review. *J. Food Sci. Technol.* **2012**, *49*, 278–293. [CrossRef] [PubMed]
6. Lynch, S.A.; Mullen, A.M.; O'Neill, E.; Drummond, L.; Alvarez, C. Opportunities and perspectives for utilisation of co-products in the meat industry. *Meat Sci.* **2018**, *144*, 62–73. [CrossRef] [PubMed]
7. Sundar, V.J.; Gnanamani, A.; Muralidharan, C.; Chandrababu, N.K.; Mandal, A.B. Recovery and utilization of proteinous wastes of leather making: A review. *Rev. Environ. Sci. Biotechnol.* **2011**, *10*, 151–163. [CrossRef]
8. Bajza, Z.; Vrucek, V. Thermal and enzymatic recovering of proteins from untanned leather waste. *Waste Manag.* **2001**, *21*, 79–84. [CrossRef]
9. Zhang, Y.H.; Olsen, K.; Grossi, A.; Otte, J. Effect of pretreatment on enzymatic hydrolysis of bovine collagen and formation of ACE-inhibitory peptides. *Food Chem.* **2013**, *141*, 2343–2354. [CrossRef] [PubMed]
10. Damrongsakkul, S.; Ratanathammapan, K.; Komolpis, K.; Tanthapanichakoon, W. Enzymatic hydrolysis of rawhide using papain and neutrase. *J. Ind. Eng. Chem.* **2008**, *14*, 202–206. [CrossRef]
11. Jian, S.; Wenyi, T.; Wuyong, C. Ultrasound-accelerated enzymatic hydrolysis of solid leather waste. *J. Clean Prod.* **2008**, *16*, 591–597. [CrossRef]
12. Anzani, C.; Prandi, B.; Tedeschi, T.; Baldinelli, C.; Sorlini, G.; Wierenga, P.A.; Dossena, A.; Sforza, S. Degradation of collagen increases nitrogen solubilisation during enzymatic hydrolysis of fleshing meat. *Waste Biomass. Valor.* **2018**, *9*, 1113–1119. [CrossRef]
13. Toldra, F.; Aristoy, M.C.; Mora, L.; Reig, M. Innovations in value-addition of edible meat by-products. *Meat Sci.* **2012**, *92*, 290–296. [CrossRef]
14. Korhonen, H.; Pihlanto, A. Bioactive peptides: Production and functionality. *Int. Dairy J.* **2006**, *16*, 945–960. [CrossRef]
15. Arihara, K.; Ohata, M. Functional meat products. In *Handbook of Meat Processing*; Toldrá, F., Ed.; Wiley-Blackwell: Ames, IA, USA, 2010; pp. 423–439.

16. Di Bernardini, R.; Harnedy, P.; Bolton, D.; Kerry, J.; O'Neill, E.; Mullen, A.M.; Hayes, M. Antioxidant and antimicrobial peptidi hydrolysates from muscle protein sources and by-products. *Food Chem.* **2011**, *124*, 1296–1307. [CrossRef]
17. Saiga, A.; Tanabe, S.; Nishimura, T. Antioxidant activity of peptides obtained from porcine myofibrillar proteins by protease treatment. *J. Agric. Food Chem.* **2003**, *51*, 3661–3667. [CrossRef]
18. Wang, B.-S.; Chang, L.-W.; Wu, H.-C.; Huang, S.-L.; Chu, H.-L.; Huang, M.-H. Antioxidant and anti-tyrosinase activity of aqueous extracts of green asparagus. *Food Chem.* **2011**, *127*, 141–146. [CrossRef]
19. Yu, Z.; Zeng, W. Antioxidant, antibrowning, and cytoprotective activities of *Ligustrum robustum* (Rxob.) Blume extract. *J. Food Sci. Technol.* **2013**, *78*, C1354–1362.
20. Bose, B.; Choudhury, H.; Tandon, P.; Kumaria, S. Studies on secondary metabolite profiling, anti-inflammatory potential in vitrophotoprotective and skin-aging related enzyme inhibitory activities of *Malaxis acuminata*, a threatened orchid of nutraceutical importance. *J. Photochem. Photobiol. B* **2017**, *173*, 686–695. [CrossRef]
21. Ferri, M.; Rondini, G.; Calabretta, M.M.; Michelini, E.; Vallini, V.; Fava, F.; Roda, A.; Minnucci, G.; Tassoni, A. White grape pomace extracts, obtained by a sequential enzymatic plus ethanol-based extraction, exert antioxidant, anti-tyrosinase and anti-inflammatory activities. *New Biotechnol.* **2017**, *39*, 51–58. [CrossRef] [PubMed]
22. Karioti, A.; Protopappa, A.; Megoulas, N.; Skaltsa, H. Identification of tyrosinase inhibitors from *Marrubium velutinum* and *Marrubium cylleneum*. *Bioorgan. Med. Chem.* **2007**, *15*, 2708–2714. [CrossRef] [PubMed]
23. Monari, S.; Ferri, M.; Russo, C.; Prandi, B.; Tedeschi, T.; Bellucci, P.; Zambrini, A.V.; Donati, E.; Tassoni, A. Enzymatic production of bioactive peptides from scotta, an exhausted by-product of ricotta cheese processing. *PLoS ONE* **2019**, *14*, e0226834. [CrossRef] [PubMed]
24. Ferri, M.; Graen-Heedfeld, J.; Bretz, K.; Guillon, F.; Michelini, E.; Calabretta, M.M.; Lamborghini, M.; Gruarin, N.; Roda, A. Kraft, A.; et al. Peptide fractions obtained from rice by-products by means of an environment-friendly process show in vitro health-related bioactivities. *PLoS ONE* **2017**, *12*, e0170954. [CrossRef]
25. Schurink, M.; van Berkel, W.; Wichers, H.; Boeriu, C. Novel peptides with tyrosinase inhibitory activity. *Peptides* **2007**, *28*, 485–495. [CrossRef]
26. Mapiye, C.; Dugan, M.E.R.; Juarez, M.; Basarab, J.A.; Baron, V.S.; Turner, T.; Yang, X.; Aldai, N.; Aalhus, J.L. Influence of alpha-tocopherol supplementation on trans-18:1 and conjugated linoleic acid profiles in beef from steers fed a barley-based diet. *Animal* **2012**, *6*, 1888–1896. [CrossRef]
27. Marzocchi, S.; Pasini, F.; Baldinelli, C.; Caboni, M.F. Value-addition of beef meat by-products: Lipid characterization by chromatographic techniques. *J. Oleo Sci.* **2018**, *67*, 143–150. [CrossRef]
28. Daley, C.A.; Abbott, A.; Doyle, P.S.; Nader, G.A.; Larson, S. A review of fatty acid profiles and antioxidant content in grass-fed and grain-fed beef. *Nutr. J.* **2010**, *9*, 10. [CrossRef]
29. AOAC International. Official Methods of Analysis. 2002. Available online: http://www.aoac.org/ (accessed on 14 December 2020).
30. Lowry, O.H.; Rosebrough, N.J.; Farr, A.L.; Randall, R.J. Protein measurement with the Folin phenol reagent. *J. Biol. Chem.* **1951**, *193*, 265–275. [CrossRef]
31. Christie, W.W. A simple procedure for rapid transmethylation of glycerolipids and cholesteryl esters. *J. Lipid Res.* **1982**, *23*, 1072–1075. [CrossRef]
32. Verardo, V.; Gomez-Caravaca, A.M.; Gori, A.; Losi, G.; Caboni, M.F. Bioactive lipids in the butter production chain from Parmigiano Reggiano cheese area. *J. Sci. Food Agr.* **2013**, *93*, 3625–3633. [CrossRef]
33. Ferri, M.; Gianotti, A.; Tassoni, A. Optimisation of assay conditions for the determination of antioxidant capacity and polyphenols in cereal food components. *J. Food Compos. Anal.* **2013**, *30*, 94–101. [CrossRef]
34. Rai, A.K.; Nived, C.; Sakhare, P.Z.; Suresh, P.V.; Bhaskar, N.; Mahendrakar, N.S. Optimization of acid hydrolysis conditions of delimed tannery fleshings by response surface method. *J. Sci. Ind. Res. India* **2009**, *68*, 967–974.
35. Bhaskar, N.; Sakhare, P.Z.; Suresh, P.V.; Gowda, L.R.; Mahendrakar, N.S. Biostabilization and preparation of protein hydrolysates from delimed leather fleshings. *J. Sci. Ind. Res. India* **2007**, *66*, 1054–1063.
36. USDA-United States Department of Agriculture. Agricultural Research Service, National Nutrient Database for Standard Reference Legacy. Available online: https://fdc.nal.usda.gov (accessed on 14 December 2020).
37. White, J.S.; White, D.C. *Source Book of Enzymes*, 1st ed.; CRC Press: Boca Raton, FL, USA, 1997.
38. Olsen, J.V.; Ong, S.E.; Mann, M. Trypsin cleaves exclusively C-terminal to arginine and lysine residues. *Mol. Cell Proteom.* **2004**, *3*, 608–614. [CrossRef]
39. Haslaniza, H.; Maskat, M.Y.; Aida, W.M.W.; Mamot, S.; Saadiah, I. Optimization of enzymatic hydrolysis of cockle (Anadara Granosa) meat wash water precipitate for the development of seafood flavor. *Int. Food Res. J.* **2013**, *20*, 3053–3059.
40. Ryder, K.; Bekhit, A.E.D.; McConnell, M.; Carne, A. Towards generation of bioactive peptides from meat industry waste proteins: Generation of peptides using commercial microbial proteases. *Food Chem.* **2016**, *208*, 42–50. [CrossRef]
41. Choi, J.S.; Moon, H.E.; Roh, M.K.; Ha, V.M.; Lee, B.B.; Cho, K.K.; Choi, I.S. Physiological properties of Scomber japonicus meat hydrolysate prepared by subcritical water hydrolysis. *J. Environ. Biol.* **2016**, *37*, 57–63.
42. Baggio, S.R.; Bragagnolo, N. The effect of heat treatment on the cholesterol oxides, cholesterol, total lipid and fatty acid contents of processed meat products. *Food Chem.* **2006**, *95*, 611–619. [CrossRef]

3. Muchenje, V.; Hugo, A.; Dzama, K.; Chimonyo, M.; Strydom, P.E.; Raats, J.G. Cholesterol levels and fatty acid profiles of beef from three cattle breeds raised on natural pasture. *J. Food Compos. Anal.* **2009**, *22*, 354–358. [CrossRef]
4. Brugiapaglia, A.; Lussiana, C.; Destefanis, G. Fatty acid profile and cholesterol content of beef at retail of Piemontese, Limousin and Friesian breeds. *Meat Sci.* **2014**, *96*, 568–573. [CrossRef] [PubMed]

foods

MDPI

Article

Microwave Assisted Extraction of Bioactive Carbohydrates from Different Morphological Parts of Alfalfa (*Medicago sativa* L.)

Daniela Alejandra Solarte [1], Ana Isabel Ruiz-Matute [2], Diana M. Chito-Trujillo [1], Maite Rada-Mendoza [1] and María Luz Sanz [2,*]

1 Grupo de Investigación Biotecnología, Calidad Medioambiental y Seguridad Agroalimentaria (BICAMSA), Universidad del Cauca, Popayán 190003, Colombia; danielasolarte@unicauca.edu.co (D.A.S.); dchito@unicauca.edu.co (D.M.C.-T.); mrada@unicauca.edu.co (M.R.-M.)

2 Instituto de Química Orgánica General (IQOG-CSIC), Juan de la Cierva 3, 28006 Madrid, Spain; ana.ruiz@csic.es

* Correspondence: mlsanz@iqog.csic.es

Abstract: Despite the nutritional properties of alfalfa, its production is mainly for animal feed and it is undervalued as a food source. In this study, the valorization of alfalfa as a potential source of bioactive carbohydrates [inositols, α-galactooligosaccharides (α-GOS)] is presented. A Box–Behnken experimental design was used to optimize the extraction of these carbohydrates from leaves, stems, and seeds of alfalfa by solid–liquid extraction (SLE) and microwave-assisted extraction (MAE). Optimal extraction temperatures were similar for both treatments (40 °C leaves, 80 °C seeds); however, SLE required longer times (32.5 and 60 min vs. 5 min). In general, under similar extraction conditions, MAE provided higher yields of inositols (up to twice) and α-GOS (up to 7 times); hence, MAE was selected for their extraction from 13 alfalfa samples. Pinitol was the most abundant inositol of leaves and stems (24.2–31.0 mg·g^{-1} and 15.5–22.5 mg·g^{-1}, respectively) while seed extracts were rich in α-GOS, mainly in stachyose (48.8–84.7 mg·g^{-1}). In addition, inositols and α-GOS concentrations of lyophilized MAE extracts were stable for up to 26 days at 50 °C. These findings demonstrate that alfalfa is a valuable source of bioactive carbohydrates and MAE a promising alternative technique to obtain functional extracts.

Keywords: alfalfa (*Medicago sativa* L.); α-galactooligosaccharides (α-GOS); inositols; sugars; microwave assisted extraction (MAE); solid-liquid extraction (SLE)

1. Introduction

Alfalfa or lucerne (*Medicago sativa*) is a perennial herbaceous plant belonging to the Fabaceae family. Worldwide, it is widely cultivated and intended for animal nutrition since it can be grown in areas with extreme abiotic factors [1] and has high protein and digestible fiber levels [2,3]. Alfalfa extracts have been proven to be an efficient dietary tool in the treatment of hypertension, metabolic disorders related to glucose, and lipid metabolism, arthritis, and kidney problems [2] and to be helpful in lowering cholesterol levels in both animals and humans [4,5]. These activities have been attributed to its content in several bioactive compounds such as sterols, triterpenes, phenolic compounds, fatty acids, and saponins [2,5–8]. While several studies have been carried out regarding these compounds, scarce attention has been paid to its bioactive carbohydrate composition [1,9].

Cyclitols (also named inositols) are minor phytochemicals with several reported biological activities (anti-diabetic, anti-inflammatory, antioxidant, and even anti-cancer properties) [1]. The presence of free inositols such as *myo*-inositol, methyl-inositols (such as pinitol and ononitol), and glycosyl-inositols (such as galactinol) have been previously detected in alfalfa leaves [9]. Alfalfa seeds contain several glycosyl-inositols and they have also been shown to be a rich source of α-galactooligosaccharides (α-GOS) from the raffinose family. α-GOS are considered prebiotics which selectively stimulate the growth and activity of a limited number of beneficial bacteria (mainly *Bifidobacterium* and *Lactobacillus*) in the colon [10–12]. Thus, alfalfa is a leguminous plant of great interest for exploitation by the food industry as a source of bioactive carbohydrates.

Extraction of bioactive carbohydrates from fruits, vegetables, and legumes have been usually carried out by conventional solid-liquid extraction (SLE) [13–16]. However, there is a growing interest in applying more efficient advanced extraction techniques which help to break vegetable tissues, releasing bioactive compounds from cell structures, as an alternative to the tedious and time-consuming SLE-based processes [17,18]. Among the most outstanding implemented techniques, Microwave Assisted Extraction (MAE) is preferred for its speed, efficiency, and safe operation [16,18–21]. During conventional treatments, heat energy is delivered through conduction–convection processes with the consequent loss of heat energy to the environment. However, during MAE, heating of the sample is produced in a targeted and selective process produced by the simultaneous combination of ionic conduction and dipole rotation which change microwave into thermal energy [22–24]. Subsequently, MAE provides shorter extraction times and more effective treatments due to the microwave properties—it can heat all samples simultaneously, without heating the vessel and with a faster energy transfer, reduce thermal gradients and unique heating selectivity, and ultimately afford better yields at lower costs [25–28].

The most relevant parameters that affect MAE treatments are solid–liquid ratio, extraction temperature, and time. Considering the high dependence among some of these parameters, MAE usually requires the optimization of the extraction conditions using experimental design approaches [29]. In closed systems, pressure is also an important variable, but it is directly dependent on temperature, it being preferable to control the latter to avoid degradation of thermolabile compounds [25]. Irradiation power can be also optimized; however, the number of extraction vessels simultaneously used mainly governs its selection. It is usually chosen as a compromise between minimizing the extraction time and avoiding solvent projections or degradation of thermolabile analytes [27]. The nature of the solvent also affects the extraction of bioactive carbohydrates; polar solvents such as water, methanol, or hydroalcoholic mixtures are commonly used. Based on their dielectric constant, water has a greater capacity to obstruct microwaves than other solvents such as methanol and, therefore, favors their penetration. On the contrary, methanol has a higher ability than water to dissipate the microwave energy as heat [30].

MAE has been mainly used for the extraction of polysaccharides (i.e., pectins, inulin, etc) from different natural matrices [20,31], although recently its utility for the extraction of bioactive low molecular weight carbohydrates (LMWC) from food residues has been addressed (i.e., lettuce leaves [18], artichoke bracts [20], and legume pods [21]). However, as far as we know, MAE efficiency for the extraction of bioactive carbohydrates from alfalfa has not yet been evaluated.

Therefore, in this work both SLE and MAE methods were optimized and compared for the effective extraction of cyclitols and α-GOS from different morphological parts of alfalfa. Thermal stability of lyophilized extracts obtained by the optimal extraction technique was also evaluated.

2. Material and Methods

2.1. Reagents

Analytical standards of pinitol, *myo*-inositol, galactinol, glucose, fructose, sucrose, raffinose, stachyose, and phenyl-β-D-glucopyranoside were purchased from Sigma Chemical Co. (St. Louis, MO, USA). Ethanol and methanol were acquired from Scharlab (Barcelona, Spain) and hydroxylamine chloride, hexamethyldisilazane and trifluoroacetic acid were purchased from Sigma Chemical Co. (St. Louis, MO, USA)

2.2. Samples

Thirteen samples from three different morphological parts: leaves (Lv1–Lv4), stems (St1–St4), and seeds (Sd1–Sd5)] of alfalfa (*Medicago sativa* L.) were purchased between June and August 2018. Leaves and stems were acquired in local markets of Nariño (Colombia), and seeds in Colombian and Spanish seed stores. The identification of samples is shown in Table S1 of Supplementary Material.

Leaves samples were dried on paper in the dark at room temperature for 72 h, while stems were cut into small pieces and dried at 40 °C in an oven for 24 h. Seeds were not subjected to any previous drying process. Samples were ground to fine particles using a domestic mill (Moulinex, Barcelona, Spain) and sieved through a 500 μm mesh. Finally, they were stored in a dry, hermetically sealed recipient protected from light until analysis at room temperature.

2.3. Extraction Methods

Lv2 and Sd2 samples were selected for the optimization of both SLE and MAE methods. Prior to the evaluation of the effect of different extraction conditions on cyclitol and α-GOS yields, optimization of solvent was tackled by SLE. Samples (0.3 g) were mixed with 10 mL of different percentages of ethanol:water and methanol:water (0–100%; 25–75%; 50–50%; 75–25%; 100–0%, *v/v*) and stirred at 75 °C for 16 min. Assays were performed in triplicate.

MAE extractions were carried out on MARS6 equipment (CEM, Matthews, NC, USA) provided with optic fiber (MTS-300, CEM, Matthews, NC, USA) for temperature control. Microwave power was set at 900 W. Samples were accurately weighed and placed along with a fixed volume of 10 mL of the selected solvent in 100 mL X-Press 1500 vessels (CEM, Matthews, NC, USA) and subjected to MAE (n = 3).

A Box–Behnken experimental design was performed in order to evaluate the effect of three independent variables (sample amount (s, g); temperature (T, °C) and time (t, min)) on extraction of cyclitols and α-GOS by SLE and MAE. These independent variables were selected as the most relevant factors according to findings in previous investigations [16,20,21]. Sample amount was varied to evaluate the effect of modifying the solid liquid ratio, using a fixed solvent volume of 10 mL (the minimum allowed by the X-Press 1500 vessels). A total of 15 experiments were carried out in random order and three central points were included to estimate the experimental error (Table 1).

Experimental ranges for factors evaluated were: T = 40, 80, 120 °C, t = 5, 32.5, 60 min, and s = 0.1, 0.3, 0.5 g. Each experiment was carried out in triplicate. Response surface methodology (RSM) was used to calculate the coefficients of R_1, the total concentration of extracted cyclitols (mg·g^{-1}) and R_2, the total concentration of α-GOS (mg·g^{-1}), both responses to be maximized, in the model proposed for leaves and seeds and to estimate the statistical significance of the regression coefficients. The quadratic model proposed was:

$$R = \beta_0 + \beta_1 T + \beta_2 t + \beta_3 s + \beta_{1,1} T^2 + \beta_{2,2} t^2 + \beta_{3,3} s^2 + \beta_{1,2} Tt + \beta_{1,3} Ts + \beta_{2,3} ts + \varepsilon \quad (1)$$

where β_0 is the intercept, β_i are the first-order coefficients, $\beta_{i,i}$ the quadratic coefficients for ith factors, $\beta_{i,j}$ the coefficients for interaction between the factors i and j and ε is the error. R_1 and R_2 were estimated by multiple linear regression (MLR) using StatGraphics Centurion XV software (Statistical Graphics Corporation, Rockville, MD, USA). A multiple response analysis, that simultaneously maximized R_1 and R_2, was also considered in seeds for selection of optimal SLE and MAE operating conditions; this function takes values between 0 (completely undesirable value) and 1 (completely desirable or ideal response).

After extraction procedure, samples were cooled in ice for 5 min and solid residue was removed by centrifugation at 4400× g at 10 °C for 10 min. A clear solution was obtained and kept at −18 °C until analysis by gas chromatography-mass spectrometry (GC-MS). Each procedure was performed in triplicate.

Table 1. Experimental Box–Behnken design and inositols, sugars, and α-GOS content obtained by SLE and MAE ($mg{\cdot}g^{-1}$). Standard deviation in brackets ($n = 3$).

No	*t* (min)	*T* (°C)	*s* (g)	Inositols ($mg{\cdot}g^{-1}$)				Sugars ($mg{\cdot}g^{-1}$)				α-GOS ($mg{\cdot}g^{-1}$)	
				Leaves		Seeds		Leaves		Seeds		Seeds	
				SLE	MAE	SLE	MAE	SLE	MAE	SLE	MAE	SLE	MAE
1	32.5	80.0	0.3	21.5 (1.0) ^{b}	29.6 (1.3) ^{a}	16.9 (0.6) ^{a}	14.7 (0.4) ^{b}	3.5 (0.1) ^{b}	4.8 (0.1) ^{a}	14.1 (0.1) ^{a}	12.7 (0.4) ^{b}	84.7 (0.7) ^{b}	107.0 (2.1) ^{a}
2	5.0	40.0	0.3	18.4 (0.8) ^{b}	**36.2 (0.6)** $^{a,\#,*}$	11.5 (0.4) ^{b}	14.6 (0.5) ^{a}	3.2 (0.1) ^{a}	**3.8 (0.1)** $^{a,\#}$	6.67 (0.2) ^{b}	12.4 (0,2) ^{a}	11.4 (0.4) ^{b}	78.6 (1.9) ^{a}
3	5.0	80.0	0.5	14.7 (0.6) ^{b}	30.9 (1.3) ^{a}	12.8 (0.3) ^{b}	**13.7(0.4)** $^{a,\#}$	3.4 (0.1) ^{a}	3.3 (0.1) ^{a}	12.5 (0.5) ^{b}	**15.1 (0.4)** $^{a,\#}$	113.7 (3.5) ^{b}	**138.4(2.6)** $^{a,\&}$
4	60.0	120.0	0.3	14.6 (0.4) ^{b}	30.8 (0.9) ^{a}	17.9 (0.4) ^{a}	16.8 (0.8) ^{a}	5.1 (0.1) ^{a}	4.8 (0.1) ^{a}	14.9 (0.3) ^{a}	13.3 (0.6) ^{b}	155.2 (6.0) ^{a}	91.4 (0.8) ^{b}
5	32.5	120.0	0.1	15.5 (0.1) ^{b}	37.5 (1.8) ^{a}	28.4 (0.4) ^{b}	29.5 (0.6) ^{a}	6.82 (0.02) ^{a}	6.7 (0.1) ^{a}	8.859 (0.4) ^{b}	9.8 (0.4) ^{a}	50.6 (2.0) ^{b}	62.4 (0.4) ^{a}
6	32.5	80.0	0.3	20.7 (1.1) ^{b}	31.7 (1.3) ^{a}	14.6 (0.4) ^{a}	14.7 (0.5) ^{a}	5.2 (0.1) ^{a}	4.9 (0.1) ^{a}	13.7 (0.2) ^{a}	11.6 (0.3) ^{b}	80.3 (1.2) ^{b}	91.1 (0.5) ^{a}
7	5.0	80.0	0.1	20.1 (0.3) ^{b}	30.9 (1.2) ^{a}	27.5 (0.4) ^{b}	29.4 (0.5) ^{a}	9.02 (0.02) ^{a}	8.4 (0.1) ^{b}	9.1 (0.3) ^{a}	9.8 (0.6) ^{a}	28.1 (0.7) ^{b}	64.0 (0.7) ^{a}
8	32.5	120.0	0.5	21.8 (0.7) ^{b}	29.7 (1.3) ^{a}	11.6 (0.4) ^{b}	14.0 (0.5) ^{a}	3.69 (0.04) ^{a}	3.4 (0.1) ^{a}	11.5 (0.5) ^{b}	16.7 (0.8) ^{a}	84.4 (0.6) ^{b}	119.5 (1.0) ^{a}
9	32.5	40.0	0.1	20.4 (0.4) ^{b}	38.4 (0.6) ^{a}	28.2 (0.5) ^{b}	29.5 (0.9) ^{a}	5.8 (0.1) ^{b}	7.0 (0.2) ^{a}	10.8 (0.4) ^{b}	12.1 (0.2) ^{a}	36.3 (0.5) ^{b}	40.6 (0.2) ^{a}
10	5.0	120.0	0.3	19.5 (0.3) ^{b}	29.1 (1.2) ^{a}	14.8 (0.3) ^{b}	17.5 (0.03) ^{a}	5.2 (0.1) ^{a}	5.0 (0.1) ^{a}	12.3 (0.3) ^{b}	15.4 (0.1) ^{a}	99.0 (1.1) ^{b}	154.1 (0.6) ^{a}
11	60.0	80.0	0.1	14.3 (0.5)^{b}	29.8 (0.2) ^{a}	31.2 (0.8) ^{a}	30.9 (0.4) ^{a}	6.79 (0.02) ^{a}	6.05 (0.01) ^{a}	16.7 (0.4) ^{a}	11.7 (0.2) ^{b}	96.1 (2.9) ^{a}	97.0 (0.8) ^{a}
12	32.5	80.0	0.3	21.1(0.02) ^{b}	29.3 (1.1) ^{a}	15.8 (0.3) ^{a}	13.9 (0.1) ^{a}	5.34 (0.03) ^{a}	5.1 (0.1) ^{a}	13.9 (0.4) ^{a}	12.5 (0.5) ^{b}	81.1 (1.6) ^{b}	106.2 (0.9) ^{a}
13	60.0	40.0	0.3	14.3 (0.4) ^{b}	26.2 (0.2) ^{a}	15.7 (0.4) ^{a}	15.7 (0.9) ^{a}	3.0 (0.1) ^{b}	5.2 (0.2) ^{a}	18.2 (0.5) ^{b}	13.5 (0.2) ^{a}	76.3 (0.4) ^{b}	101.0 (2.3) ^{a}
14	60.0	80.0	0.5	16.8 (0.8) ^{b}	27.3 (1.0) ^{a}	**12.5 (0.2)** $^{b,\&}$	13.6 (0.3) ^{a}	3.72 (0.04) ^{a}	3.25 (0.04) ^{a}	**13.5 (0.5)** $^{b,\&}$	14.6 (0.6) ^{a}	**190.8 (2.1)** $^{b,\#}$	270.6 (2.4) ^{a}
15	32.5	40.0	0.5	**31.4 (0.4)** $^{a,\&}$	28.5 (1.1) ^{b}	12.2 (0.4) ^{b}	13.5 (0.2) ^{a}	**4.4 (0.1)** $^{a,\#}$	4.1 (0.1) ^{a}	11.8 (0.5) ^{b}	17.4 (0.6) ^{a}	156.6 (0.8) ^{b}	186.0 (1.7) ^{a}

^{a,b} Different letters indicate significant differences ($p < 0.05$) for the same experiment, analyte (inositols, sugars, and α-GOS) and sample (Lv2 and Sd2), $^{\#,\&}$ Different symbols indicate significant differences ($p < 0.05$) for the same analyte (inositols, sugars, and α-GOS) and sample (Lv2 and Sd2) under optimal extraction conditions for each technique. * Optimal conditions for each treatment are marked in bold. *t*, time; *T*, temperature; *s*, sample amount; SLE, solid-liquid extraction; MAE, Microwave Assisted Extraction; α-GOS, α-galactooligosaccharides

2.4. GC-MS Analysis

Prior to the GC-MS analysis of inositols and α-GOS the alfalfa extracts were derivatized to form their corresponding trimethylsilyl oximes (TMSO) according to Ruiz-Aceituno et al. [20]. Extracts (0.5 mL) and 0.1 mL of internal standard (phenyl-β-D-glucopyranoside, 1 mg·mL^{-1}) were dried under vacuum in a miVac concentrator (Inycom, Madrid, Spain) at 40 °C. Oximes were prepared by addition of 350 μL of a 2.5% hydroxylamine chloride in pyridine solution heating at 75 °C for 30 min. Then, trimethylsilyl derivatives were obtained by addition of 350 μL of hexamethyldisilazane and 35 μL of trifluoroacetic acid at 45 °C for 30 min. Samples were centrifuged at 4400× *g* for 10 min and the supernatant was recovered.

GC-MS analysis was carried out using a 6890 N gas chromatograph coupled to a 5973 N quadruple mass detector (Agilent Technologies, Santa Clara, CA, USA) with a high-temperature capillary column HT-5 coated with 5% phenyl polycarborane-siloxane (25 m × 0.22 mm id, 0.10 μm film thickness; Analytical Science, Santa Clara, CA, USA) and using helium at 0.8 mL·min^{-1} as carrier gas. The oven temperature was programmed as follows: 160 °C (10 min), then at 10 °C·min^{-1} to 380 °C, which was maintained for 5 min. Injections (1 μL) were carried out in split mode (1:20) at 280 °C. Mass spectrometer transfer line and ion source were set at 280 °C and 230 °C, respectively. Mass spectra were acquired in electron impact mode at 70 eV, scanning in a mass range: 40–700 *m/z*. Data acquisition was conducted using HP ChemStation software (Agilent Technologies, Santa Clara, CA, USA).

Identification of cyclitols and α-GOS was carried out by comparison of their corresponding linear retention indices (I^T) and retention times and mass spectra with those of corresponding standards and data reported in the literature [13]. When analytical standards were not available, compounds were tentatively identified based on their chromatographic retention and mass spectral data. Quantitative data were determined by the internal standard method. Standard solutions of target compounds over the expected concentration range in sample extracts (sugars and inositols from 0.005 to 0.5 mg·mL^{-1}, and stachyose from 0.0025 to 1.0 mg·mL^{-1}) were used to calculate the response factors relative to the internal standard (n = 3). Ononitol, trehalose, digalactosyl-inositol, and galactose were quantified from external calibration curve of pinitol, sucrose, galactinol, and glucose, respectively.

2.5. Thermal Stability

Aliquots of 1 mL of Lv2 and Sd2 extracts obtained by MAE were lyophilized in a Cryodos −80 freeze dryer (Telstar S.A., Madrid, Spain) at −80 °C and 0.080 mbar and then, subjected to accelerated temperature conditions at 50 °C in a convection oven for 26 days, following the process described by Rodríguez-Sánchez et al. [32]. Samples from each extract were collected at 0, 5, 12, 19, and 26 days, reconstituted with Milli-Q water (1 mL) and analyzed by GC-MS, as described in previous section.

2.6. Statistical Analysis

The analysis of variance ANOVA was used to determine significant differences ($p \leq 0.05$) between different samples, using the StatGraphics Centurion XV software (Statistical Graphics Corporation, Rockville, MD, USA).

3. Results and Discussion

3.1. Bioactive Carbohydrates Composition of Alfalfa Leaves, Stems and Seeds

Chromatographic profiles of Lv2, St2, and Sd2 extracts obtained by SLE at 75 °C for 16 min using 0.3 g of sample and 10 mL of Milli-Q water are shown in Figure 1; carbohydrate composition of Lv2, St2, and Sd2 extracts, indicating their t_R and experimental I^T is detailed in Table S2 of Supplementary Material. In general, similar GC profiles were observed for leaves and stems extracts (Figure 1A,B), while those of seeds (Figure 1C) showed a different carbohydrate composition. Several free inositols such as pinitol, ononitol and *myo*-inositol were detected in leaves and stems, while seeds were a richer source of glycosyl-cyclitols such as galactosyl-pinitols, galactinol, and digalactosyl-cyclitols, as previously reported by Horbowicz et al. [9]. Other LMWC such as fructose, glucose, galactose, sucrose, and trehalose were also present in leaves and stems. Sucrose was also found in seeds extracts together with the α-GOS raffinose and stachyose. Galactinol has been described to be the main galactosyl donor implicated in α-GOS biosynthetic pathway, where stachyose is synthesized from raffinose, and this trisaccharide from sucrose [33]. These oligosaccharides, which were not detected in leaves and stems, have been related to maturation, desiccation tolerance, and storability of vegetable seeds, reducing damage caused by oxidative stress or ageing [9,33]. Considering the similar composition of leaf and stem extracts, only Lv2 and Sd2 samples were chosen for the subsequent analyses.

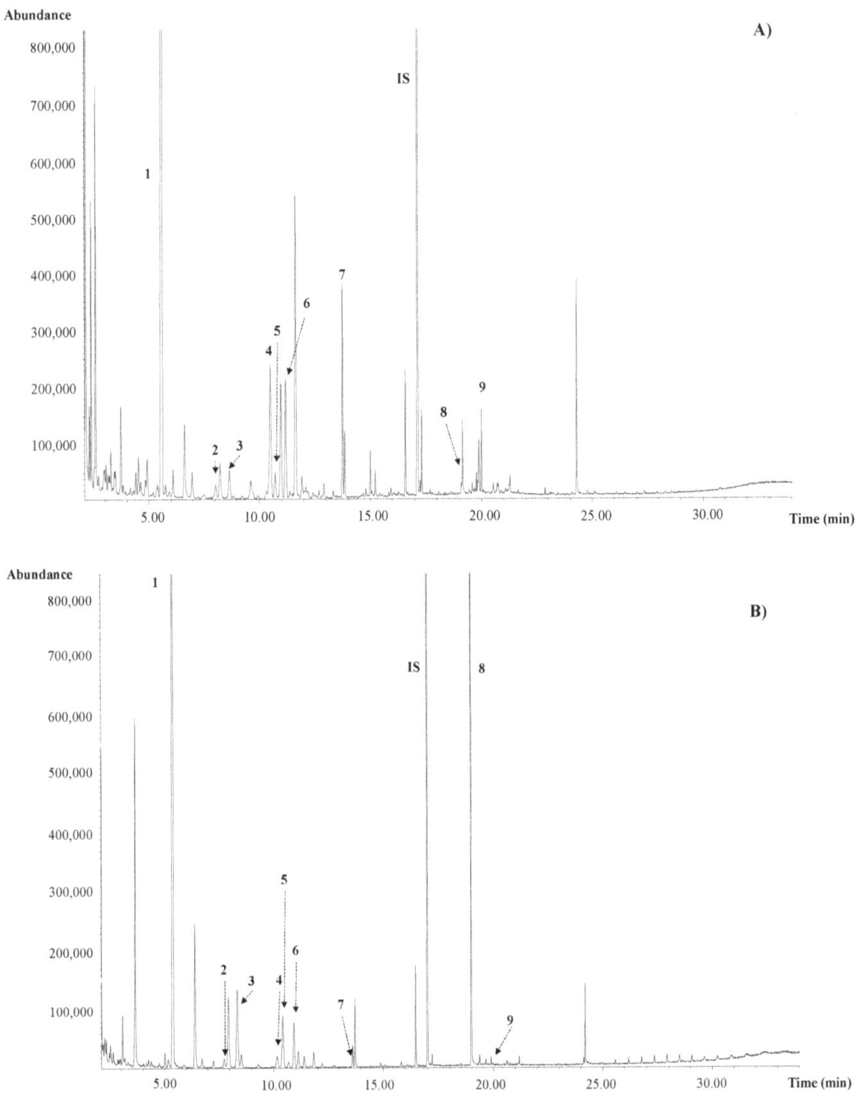

Figure 1. *Cont.*

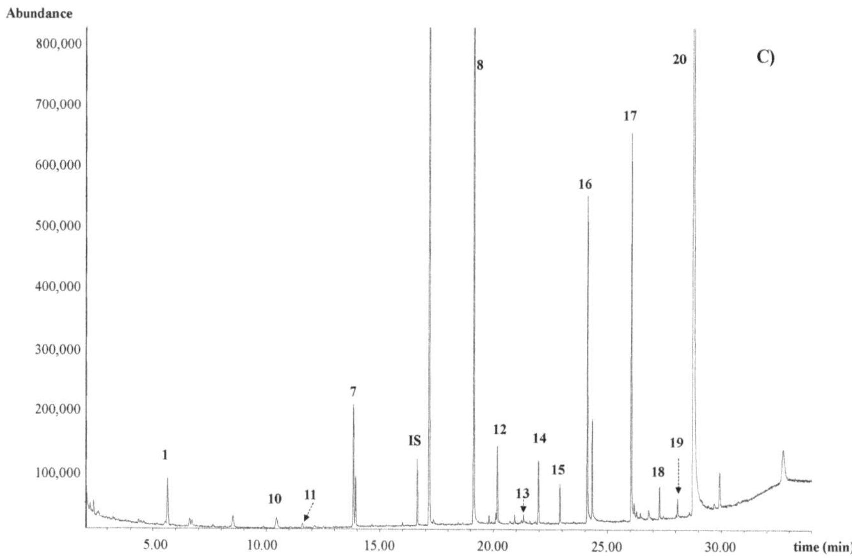

Figure 1. GC-MS profile of alfalfa SLE extracts previously derivatized to TMS-Oximes. (**A**) Leaves (Lv2); (**B**) Stems (St2); (**C**) Seeds (Sd2): 1: pinitol; 2: fructose 1; 3: fructose 2; 4: ononitol 5: glucose *E*; 6: glucose *Z*; 7: *myo*-inositol; 8: sucrose; 9: trehalose; 10: galactose *E*; 11: galactose *Z*; 12: galactosyl-pinitol A; 13: galactosyl-pinitol B; 14: galactinol; 15: digalactosyl-glycerol, 16: raffinose; 17–19: digalactosyl-inositol; 20: stachyose; IS: internal standard.

3.2. Optimization of SLE Method

Firstly, different solvents for the extraction of bioactive carbohydrates from alfalfa leaves (Lv2) and seeds (Sd2) by SLE were evaluated. This is a crucial step for the optimization of an extraction method since it has environmental, social, and economic implications. The use of GRAS (generally recognized as safe) solvents instead of contaminant and hazardous solvents is encouraged. Moreover, solvent price and recyclability are other parameters to be considered [34]. Polar solvents are the most appropriate for the extraction of carbohydrates, considering their high polarity. Moreover, these solvents have further application in MAE since they strongly absorb microwave energy due to the presence of permanent dipoles [17]. Then, water and different percentages of ethanol:water and methanol:water were assayed within an extraction cycle of 16 min at 75 °C under stirring conditions. Figure 2 depicts the carbohydrate yields extracted from leaves and seeds using these solvents. To simplify data, carbohydrates have been grouped in inositols, α-GOS and non-bioactive sugars (glucose, galactose, fructose and sucrose). Regarding inositols, no noticeable differences were observed among the different percentages of methanol:water and ethanol:water assayed for both leaves and seeds. However, yields obtained with aqueous extracts (100% water) were higher than those of alcoholic mixtures. Moreover, water extracts also presented the greatest yields of α-GOS in alfalfa seeds (up to 11% higher than yields achieved using alcohol:water mixtures). These results are in good agreement with those reported by López-Molina et al. [35] and Ruiz-Aceituno et al. [14], for the SLE of prebiotics from artichoke and inositols from pine nuts, respectively. Furthermore, from a technical point of view, the choice of water as the extraction solvent is considered the most appropriate due to its availability at low cost, its versatility in different environmental conditions and its properties as a green solvent [34,36]. On the contrary, in a study of Carrero-Carralero et al. [16], although water was the most effective solvent for the extraction of inositols and prebiotics from mung beans, it was not selected due to the high concentrations of non-bioactive sugars that were also co-extracted. This was not the case for alfalfa extractions, in which similar extraction yields of non-bioactive sugars were obtained using the different evaluated solvents for both leaves and seeds. Considering the relative low concentrations of sugars in alfalfa compared with those of bioactive carbohydrates, water was finally selected as a solvent for further studies.

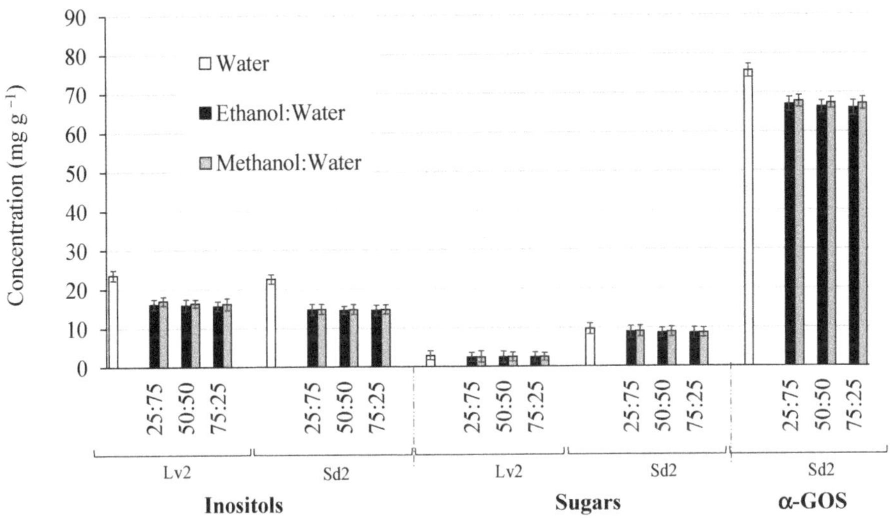

Figure 2. Inositol, sugar and α-GOS yields of Lv2 and Sd2 SLE extracts obtained using different extraction solvents. Lv2, leaves; Sd2, seeds; α-GOS, α-galactooligosaccharides

After selection of the solvent, a complete optimization of SLE was carried out. A total of 15 experiments were done in randomized order according to a Box–Behnken design to study the impact of the three independent variables (extraction time (t), temperature (T) and sample amount (s) for a fixed solvent volume of 10 mL) on the efficiency of the SLE of total inositol (R_1, mg·g^{-1}) concentrations in alfalfa leaves and of both total inositol and total α-GOS (R_2, mg·g^{-1}) in alfalfa seeds. Sugar concentrations were also considered in order to control a potential and undesirable substantial increase in their concentration under certain extraction conditions. Table 1 shows the obtained yields at the different evaluated conditions. While a variation of 2.7 times was found for inositol and sugar concentrations of Sd2 depending on the extraction conditions (inositols between 11.5 and 31.2 mg·g^{-1} and sugars between 6.6 and 18.2 mg·g^{-1}), concentrations of α-GOS underwent a variation of almost 16 times (between 11.4–190.8 mg·g^{-1}). Regarding leaves extract, a slight variation was observed for inositol concentrations (most extracts presented values between 14.3 and 21.8 mg g^{-1}, except for experiment 15 which showed 31.4 mg g^{-1}), while sugar concentrations varied from 2.98 to 9.0 mg·g^{-1}. In general, at higher temperatures and extraction times, higher inositol and α-GOS concentrations were obtained.

Response surface methodology was followed to calculate the coefficients of R_1 and R_2 in the proposed quadratic model and to estimate the statistical significance of the regression coefficients. Regarding inositols (R_1), t^2 was the most significant factor at a 95% confidence level ($p < 0.05$) for Lv2 extract, while s, s^2 and t were the factors statistically significant ($p < 0.05$) for Sd2 extract. Regarding α-GOS (R_2) for Sd2 extract, s and t were the most significant factors. The quadratic regression equations (from Equation (1)) for R_1 and R_2 are shown in Table 2. The surface plots to maximize R_1 in leaves and R_1 and R_2 in seeds after excluding no significant ($p > 0.05$) terms in the model are shown in

Figure S1A–C of Supplementary Material. Good fits were found for both inositols and α-GOS in Sd2 extract, where the R^2 were 0.97 and 0.67, respectively. The resultant optimal conditions which maximized inositols and α-GOS yields were 80 °C, 60 min, 0.1 g, and 80 °C, 60 min, 0.5 g, respectively. On the contrary, a low fit quality of the proposed model was found for inositols in Lv2 (R^2 = 0.58), probably due to the low variability in their concentrations among experiments. Then, SLE conditions for extraction of inositols from Lv2 were selected considering the experiment that provided the highest inositol yields from those evaluated (Table 1, experiment 15: 40 °C, 32.5 min and 0.5 g).

Table 2. Regression equations to maximize the extraction of inositols (R_1) and α-GOS (R_2) from leaves (Lv2) and seeds (Sd2) by solid-liquid extraction (SLE) and microwave assisted extraction (MAE).

		SLE		MAE	
	Response Variable	**Model Equation**	R^2	**Model Equation**	R^2
Leaves	R_1	$R_1 = 4.174 + 0.066{*}T + 0.566{*}t + 52.000{*}s - 0.459{*}Ts - 0.087{*}t^2$	57.8	$R_1 = 43.614 - 0.086{*}T - 0.270{*}t - 12.625{*}s + 0.003{*}Tt$	59.2
Seeds	R_1	$R_1 = 38.005 + 0.049{*}t - 120.411{*}s + 131.518{*}s^2$	97.2	$R_1 = 42.741 - 0.082{*}t - 136.683{*}s + 0.001{*}t^2 + 160.721{*}s^2$	98.8
	R_2	$R_2 = -12.385 + 1.210{*}t + 209.000{*}s$	66.4	$R_2 = 54.903 - 0.785{*}t + 135.017{*}s + 4.509{*}ts$	65.7

Finally, a multiple response analysis that simultaneously maximized the extraction of inositols and α-GOS (R_1 and R_2) in Sd2 extract was carried out. Optimal conditions were 80 °C, 60 min, and 0.5 g.

3.3. *Optimization of MAE Method*

Water was selected as the extraction solvent for MAE experiments, considering the results obtained during the SLE treatment of alfalfa leaves and seeds, its high capacity to obstruct microwaves which favor their penetration [30], and its green nature. Similar to SLE procedure, optimization of MAE conditions (T, t and s) was carried out following a Box–Behnken design. Table 1 shows the experimental results of inositol and α-GOS concentrations obtained for the 15 assays performed. A slight variation of inositol concentrations depending on the extraction conditions applied was also found for Lv2 sample (between 26.2 and 38.4 mg·g^{-1}), while changes in inositol and α-GOS concentrations of Sd2 were greater (between 13.5 and 30.9 mg·g^{-1} and between 40.6 and 270.6 mg·g^{-1}, respectively). In general, under similar time and temperature conditions, inositol extraction yields in Sd2 were higher in those experiments carried out using lower sample amounts (e.g., experiment 9: 29.5 mg·g^{-1} vs. experiment 15: 13.5 mg·g^{-1}). On the contrary, similar extraction yields were obtained by varying the time (for the same temperature and quantity of sample), probably due to the fact that 5 min was enough to achieve the maximum concentration of inositols under these conditions. Regarding α-GOS, higher concentrations were obtained when temperature increased. As previously reported by Alexandru et al. [28], at higher temperatures, solvent viscosity decreased which enhances the diffusivity and then the extraction efficiency. Regarding sugars, they varied from 3.25 to 8.36 mg·g^{-1} in Lv2 and from 9.78 to 17.39 mg·g^{-1} in Sd2.

However, it is difficult to evaluate the influence of extraction parameters independently; thus, response surface methodology was used to calculate the coefficients of R_1 and R_2 and to estimate the statistical significance of the regression coefficients. In this case, s and Tt were the most statistically significant factors ($p < 0.05$) for maximization of R_1 in Lv2 sample, while s and s^2 were those for maximization of R_1, and s and ts for R_2 in Sd2 extract. The quadratic equations, after excluding non-significant parameters ($p > 0.05$), can be observed in Table 2 (response surface plots are shown in Figure S1D–F of Supplementary Material). According to SLE results, low fit quality of the quadratic model was found for inositols in Lv2 extract (R^2 = 59%), probably considering the low variability of the total inositols detected under different experimental conditions. Considering the low efficiency of the model, the experiment 2 (40 °C, 5 min, and 0.3 g) was selected as MAE optimal conditions taking into account that it provided high inositol yields (Table 1), minimizing the temperature and extraction time, and therefore the energy consumption.

The quadratic model for Sd2 sample appropriately described the variability of R_1 and R_2 (R^2 = 99% and 66%, respectively). Then, the optimal conditions for maximization of inositols were: 80 °C, 5 min, 0.1 g, and for α-GOS were: 80 °C, 60 min, 0.5 g. The multiple response analysis performed to maximize the extraction of inositols and α-GOS (R_1 and R_2) by MAE displayed as optimal conditions: 80 °C, 5 min, and 0.5 g. Thus, these conditions were selected for further experiments with alfalfa seeds.

3.4. Comparison of SLE vs. MAE

The efficiency of extraction method could be established by comparison of bioactive inositols, α-GOS, and other sugar yields obtained at similar experimental conditions assayed (Table 1). Inositol concentrations obtained by MAE were, in general, higher than those obtained by SLE in both Lv2 and Sd2, A similar behavior was also observed for α-GOS concentration in seeds. Regarding sugar content, non-significant differences were found for leaves samples by both techniques, contrarily to that observed for seeds. Nevertheless, these concentrations were lower than those of inositols and α-GOS in most cases. Then, the implementation of an additional fractionation step for the subsequent removal of sugars was not considered mandatory.

Regarding the extraction of bioactive carbohydrates under optimal conditions, MAE required shorter extraction times for both leaves and seeds (MAE 5 min vs. SLE 32.5 min Lv2 and 60 min Sd2) and smaller amounts of samples for leaves (MAE 0.3 g vs. SLE 0.5 g) than SLE. Similar temperatures were selected as optimal for both treatments (80 °C for seeds and 40 °C for leaves). Under these optimal conditions, significant higher inositol yields in both leaves and seeds were obtained by MAE (see concentration values marked in bold in Table 1), however, greater concentrations of α-GOS were obtained by SLE.

From these results, although both techniques could be appropriate to extract bioactive carbohydrates from alfalfa, considering the advantages of MAE in terms of extraction times vs. inositols and α-GOS yields, this technique was selected for further experiments. These results were in good agreement with those reported by Carrero-Carralero et al. [16], who found that MAE provided similar bioactive carbohydrates yields than SLE, but with shorter extraction times. Similar results were also observed by Zuluaga et al. [21] for the extraction of inositols from legume pods.

3.5. Analysis of Bioactive Carbohydrates of Alfalfa Samples

Optimal MAE conditions were applied to obtain extracts enriched in inositols and α-GOS from leaves (Lv1, Lv2, Lv3, and Lv4) and seeds samples (Sd1, Sd2, Sd3, Sd4, and Sd5). Extraction conditions of leaves were also used for stems (St1, St2, St3 and St4) considering their similar composition. Table 3 shows the concentrations of bioactive carbohydrates and sugars present in the analyzed samples. Leaves and stems were mainly constituted by inositols (total concentrations between 27.0 and 37.0 mg·g^{-1} for leaves and 17.9 and 24.9 mg·g^{-1} for stems). Pinitol was the most abundant inositol (24.2–31.0 mg·g^{-1} in leaves and 15.5–22.5 mg·g^{-1} in stems), mainly in Lv1 and Lv2. In general, similar concentrations of ononitol and *myo*-inositol were detected in all leaves and stems samples. Pinitol, ononitol, and *myo*-inositol concentrations were similar to that reported by Horbowicz et al. [9] and higher than those reported by Al-Suod et al. [15] for alfalfa leaves and stems after soxhlet ethanolic extraction. Moreover, these last authors did not detect ononitol in these samples, although they reported the presence of D-*chiro* and *scyllo*-inositols, non-identified in the present work. Regarding sugars, low concentrations were found for glucose, fructose, and sucrose in all leaves and stems samples. While similar concentrations of glucose and fructose were extracted by Horbowicz et al. [9] from old leaves using ethanol:water (1:1, *v/v*) at 80 °C, higher sucrose contents were found by these authors.

Table 3. Inositol, α-GOS and other low molecular weight carbohydrate content (mg·g^{-1}) of MAE leaves, stems and seeds alfalfa extracts. Standard deviation in brackets ($n = 3$).

	Leaves				Stems				Seeds				
	Lv1	**Lv2**	**Lv3**	**Lv4**	**St1**	**St2**	**St3**	**St4**	**Sd1**	**Sd2**	**Sd3**	**Sd4**	**Sd5**
Pinitol	31.0 ^{a} (1.4)	30.1 ^{a} (0.4)	24.3 ^{b} (0.6)	24.2 ^{b} (0.1)	19.3 ^{b} (1.2)	22.5 ^{a} (0.9)	15.5 ^{c} (0.8)	19.5 ^{b} (0.6)	2.3 ^{c} (0.1)	2.5 ^{b} (0.1)	2.7 ^{a} (0.1)	2.3 ^{c} (0.1)	2.5 ^{b} (0.1)
Fructose	0.7 ^{d} (0.1)	0.9 ^{c} (0.1)	1.7 ^{a} (0.1)	1.5 ^{b} (0.1)	1.1 ^{d} (0.1)	3.0 ^{a} (0.2)	1.4 ^{c} (0.1)	1.8 ^{b} (0.2)	-	-	-	-	-
Ononitol	1.0 ^{c} (0.1)	4.6 ^{a} (0.1)	1.6 ^{b} (0.1)	1.1 ^{c} (0.1)	1.1 ^{b} (0.1)	1.4 ^{a} (0.1)	1.4 ^{a} (0.1)	1.0 ^{b} (0.1)	-	-	-	-	-
Glucose	0.9 ^{d} (0.1)	2.4 ^{a} (0.1)	1.2 ^{b} (0.1)	1.0 ^{c} (0.1)	1.2 ^{c} (0.1)	2.4 ^{a} (0.1)	1.5 ^{b} (0.1)	1.1 ^{c} (0.1)	-	-	-	-	-
***myo*-Inositol**	1.1 ^{c} (0.1)	2.3 ^{a} (0.1)	1.6 ^{b} (0.1)	1.7 ^{b} (0.1)	1.0 ^{a} (0.1)	1.0 ^{a} (0.1)	1.0 ^{a} (0.1)	1.1 ^{a} (0.1)	2.7 ^{b} (0.1)	3.1 ^{a} (0.1)	3.2 ^{a} (0.1)	2.8 ^{b} (0.1)	3.1 ^{a} (0.1)
Sucrose	-	-	1.9 ^{a} (0.1)	1.7 ^{b} (0.1)	2.8 ^{a} (0.1)	2.6 ^{b} (0.1)	1.7 ^{d} (0.1)	2.4 ^{c} (0.1)	11.7 ^{b} (0.1)	10.5 ^{d} (0.1)	11.0 ^{c} (0.4)	12.5 ^{a} (0.1)	12.5 ^{a} (0.2)
Galactinol	-	-	-	-	-	-	-	-	5.2 ^{c,d} (0.1)	5.3 ^{b,c} (0.3)	5.5 ^{a} (0.1)	5.1 ^{d} (0.1)	5.4 ^{a,b} (0.1)
Digalactosyl-inositol	-	-	-	-	-	-	-	-	5.1 ^{c} (0.1)	6.4 ^{a} (0.4)	5.3 ^{b,c} (0.1)	5.2 ^{c} (0.1)	5.4 ^{b} (0.2)
Raffinose	-	-	-	-	-	-	-	-	7.9 ^{c} (0.1)	8.4 ^{a} (0.1)	8.4 ^{a} (0.1)	7.6 ^{d} (0.1)	8.0 ^{b} (0.1)
Digalactosyl-inositol	-	-	-	-	-	-	-	-	5.9 ^{c} (0.2)	7.0 ^{a} (0.1)	7.0 ^{a} (0.1)	5.9 ^{c} (0.1)	6.9 ^{b} (0.1)
Digalactosyl-inositol	-	-	-	-	-	-	-	-	-	5.2 ^{a,b} (0.1)	5.2 ^{a,b} (0.1)	5.1 ^{b} (0.1)	5.3 ^{a} (0.1)
Stachyose	-	-	-	-	-	-	-	-	48.8 ^{e} (0.6)	74.8 ^{b} (0.9)	84.7 ^{a} (5.7)	49.8 ^{d} (0.4)	70.0 ^{c} (1.6)

$^{a–d}$ Different letters indicate significant differences ($p < 0.05$) for samples of each morphological part of alfalfa. - Non-detected.

Seed extracts were rich in α-GOS (56.7–93.1 mg·g^{-1}), stachyose was the most abundant (48.8–84.7 mg·g^{-1}), mainly in Sd3 extract. Significant differences were observed in stachyose concentrations among the different seed samples. However, these values (mainly in Sd1 and Sd4) were similar to those reported by Horbowicz et al. [9]; samples Sd2, Sd3, and Sd5 presented higher concentrations. Pinitol (2.3–2.7 mg·g^{-1}) and galactinol (5.1–5.5 mg·g^{-1}) contents were also higher than those reported by Horbowicz et al. [9]. These higher yields of bioactive carbohydrates found in this work could be due to the higher efficiency of the developed MAE method in comparison with the SLE method used by Horbowicz et al. [9].

3.6. Stability Study

In order to evaluate the thermal stability of obtained extracts due to their potential use as food ingredient, Lv2 and Sd2 extracts were lyophilized and stored at 50 °C for 26 days. The variability of inositols and α-GOS concentrations over time is shown in Table 4. Contents of inositols in both Lv2 and Sd2 extracts showed a slight decrease during storage time (from 36.8 to 34.5 mg·g^{-1} and from 28.3 to 27.2 mg·g^{-1}, respectively). Similarly, α-GOS concentrations decreased less than 5% after 26 days (from 83.2 mg·g^{-1} to 79.1 mg·g^{-1}). On the contrary a marked decrease was observed in sugar concentrations of these samples (between 19.7% and 24.4%); this significant decrease could be due to the degradation of these carbohydrates to give rise to organic acids, hydroxymethylfurfural, and unwanted colored compounds, among others [37,38]. A similar behavior (stability of inositols and decrease of sugars) has been previously observed during storage of other matrices (e.g., dried fruits, [39]). Stability of free inositol has been attributed to the absence of non-reducing groups in their molecule, which seems to confer stability during thermal treatments [40]. Alterations experimented by bioactive carbohydrates during storage were not drastic, so these extracts had an acceptable stability to be used as food ingredients.

Table 4. Inositol, α-GOS, and sugar concentrations (mg g^{-1}) of MAE seeds (Sd2) and leaves (Lv2) lyophilized extracts stored at 50 °C for different times. Standard deviation in brackets (n = 3).

	Lv2		Sd2		
Sampling Time (days)	**Inositols**	**Sugars**	**Inositols**	**α-GOS**	**Sugars**
0	36.8 (0.6) [a]	3.5 (0.07) [a]	28.3 (0.5) [a]	83.2 (1.0) [a]	10.5 (0.2) [a]
5	35.8 (0.2) [b]	3.16 (0.02) [b]	27.9 (0.4) [a,b]	80.8 (1.3) [b]	8.2 (0.1) [b]
12	35.0 (0.1) [b,c]	3.07 (0.08) [b,c]	27.4 (0.3) [b,c]	79.8 (1.0) [b,c]	8.0 (0.1) [b]
19	34.8 (0.2) [c,d]	2.98 (0.02) [c]	27.3 (0.2) [c]	79.3 (0.5) [c]	8.0 (0.2) [b]
26	34.5 (0.1) [d]	2.81 (0.01) [c]	27.2 (0.1) [c]	79.1(0.5) [c]	7.9 (0.1) [b]

[a–d] Different letters indicate significant differences ($p < 0.05$) for each compound at the different storage times. α-GOS, α-galactooligosaccharides.

4. Conclusions

Optimized SLE and MAE methods to obtain extracts rich in inositols and α-GOS from different morphological parts of alfalfa (leaves, stems, and seeds) are proposed. Although both techniques were appropriate to extract bioactive carbohydrates from alfalfa using water as GRAS solvent, MAE presented advantages in terms of extraction times vs. inositols and α-GOS yields. Moreover, in the case of leaves, MAE also provided similar inositol concentrations to the classical treatment from less raw material. The increased speed and the efficiency of the extraction process by microwaves is advantageous considering the lower consumption of energy and, thus, the costs, which should be taken into account for the scaling up of the extraction process. Thus, the MAE methodology here proposed could be considered a promising efficient and green strategy to obtain bioactive carbohydrate extracts to be used as functional food ingredients. Moreover, this work opens new routes of revalorization of alfalfa (leaves, stems, and seeds) as a potential source of bioactive carbohydrates, such as inositols and α-GOS, of interest for the agri-food industry.

Supplementary Materials: The following are available online at https://www.mdpi.com/article/10.3390/foods10020346/s1, Table S1: Origin and identification of samples; Table S2: Cyclitols, α-GOS and other sugars identified in alfalfa leaves, stems and seeds extracts obtained by SLE at 75 °C for 16 min using 0.3 g of sample and 10 mL of Milli-Q water. Figure S1: Response surface plots for SLE of A. inositols from Lv2; B. inositols from Sd2; C. α-GOS from Sd2, and for MAE of D. inositols from Lv2, E. inositols from Sd2, and F. α-GOS from Sd2.

Author Contributions: Conceptualization, M.L.S. and M.R.-M.; Methodology, D.M.C.-T. and A.I.R.-M.; Software, D.A.S., M.L.S.; Formal Analysis, D.A.S., A.I.R.-M.; Investigation, D.A.S., M.L.S., M.R.-M., D.M.C.-T., A.I.R.-M.; Resources, M.L.S., M.R.-M.; Writing—Original Draft Preparation, D.A.S., D.M.C.-T.; Writing—Review & Editing, A.I.R.-M., M.R.-M.; Supervision, M.L.S., M.R.-M.; Project Administration, M.L.S., M.R.-M.; Funding Acquisition, M.L.S., D.M.C.-T., M.R.-M. All authors have read and agreed to the published version of the manuscript.

Funding: Authors thank the Spanish Ministry of Economy, Industry and Competitiveness (project AGL2016-80475-R, AEI/FEDER, UE), Ministry of Science and Innovation and Spanish State Research Agency (project reference PID2019-106405GB-I00/AEI/10.13039/501100011033), the Comunidad of Madrid and European funding from FSE and FEDER programs (project reference S2018/BAA-4393, AVANSECAL-II-CM) and to Universidad del Cauca (501100005682) of Colombia (project ID-4652) for financial support.

Conflicts of Interest: The authors declare no conflict of interest.

References

1. Ratiu, I.A.; Al-Suod, H.; Ligor, M.; Monedeiro, F.; Buszewski, B. Effects of growth conditions and cultivability on the content of cyclitols in *Medicago sativa*. *Int. J. Environm. Sci.Technol.* **2020**, *18*, 33–48. [CrossRef]
2. Martínez, R.; Kapravelou, G.; Porres, J.M.; Melesio, A.M.; Heras, L.; Cantarero, S.; Gribblec, F.M.; Parker, H.; Aranda, P.; López-Jurado, M. *Medicago sativa* L., a functional food to relieve hypertension and metabolic disorders in a spontaneously hypertensive rat model. *J. Funct. Foods* **2016**, *26*, 470–484. [CrossRef]
3. Sahni, P.; Sharma, S. Influence of processing treatments on cooking quality, functional properties, antinutrients, bioactive potenti and mineral profile of alfalfa. *LWT* **2020**, *132*, 109890. [CrossRef]
4. Hong, Y.H.; Chao, W.W.; Chen, M.L.; Lin, B.F. Ethyl acetate extracts of alfalfa (*Medicago sativa* L.) sprouts inhibit lipopolysaccharid induced inflammation *in vitro* and *in vivo*. *J. Biomed. Sci.* **2009**, *16*, 64. [CrossRef] [PubMed]
5. Silva, L.R.; Pereira, M.J.; Azevedo, J.; Gonçalves, R.F.; Valentão, P.; de Pinho, P.G.; Andrade, P.G. *Glycine max* (L.) Merr., *Vigna radia* L. and *Medicago sativa* L. sprouts: A natural source of bioactive compounds. *Food Res. Int.* **2013**, *50*, 167–175. [CrossRef]
6. Bora, K.S.; Sharma, A. Phytochemical and pharmacological potential of *Medicago sativa*: A review. *Pharm. Biol.* **2011**, *49*, 211–22 [CrossRef] [PubMed]
7. Liu, X.G.; Huang, M.Y.; Gao, P.Y.; Liu, C.F.; Sun, Y.Q.; Lv, M.C.; Yaod, G.D.; Zhang, L.X.; Li, D.Q. Bioactive constituents from *Medicago sativa* L. with antioxidant, neuroprotective and acetylcholinesterase inhibitory activities. *J. Funct. Foods.* **2018**, *4* 371–380. [CrossRef]
8. Wrona, O.; Rafińska, K.; Walczak-Skierska, J.; Mozeński, C.; Buszewski, B. Extraction and Determination of Polar Bioactiv Compounds from Alfalfa (*Medicago sativa* L.) Using Supercritical Techniques. *Molecules* **2019**, *24*, 4608. [CrossRef]
9. Horbowicz, M.; Obendorf, R.L.; McKersie, B.D.; Viands, D.R. Soluble saccharides and cyclitols in alfalfa (*Medicago sativa* L somatic embryos, leaflets, and mature seeds. *Plant Sci.* **1995**, *109*, 191–198. [CrossRef]
10. Marín-Manzano, M.C.; Hernandez-Hernandez, O.; Diez-Municio, M.; Delgado-Andrade, C.; Moreno, F.J.; Clemente, A. Prebiot properties of non-fructosylated α-galactooligosaccharides from pea (*Pisum sativum* L.) using infant fecal slurries. *Foods* **202** *9*, 921. [CrossRef]
11. den Ende, W.V. Multifunctional fructans and raffinose familyoligosaccharides. *Front. Plant Sci.* **2013**, *4*, 247.
12. Davani-Davari, D.; Negahdaripour, M.; Karimzadeh, I.; Seifan, M.; Mohkam, M.; Masoumi, S.J.; Berenjian, A.; Ghasemi, Prebiotics: Definition, Types, Sources, Mechanisms, and Clinical Applications. *Foods* **2019**, *8*, 92. [CrossRef]
13. Hernández-Hernández, O.; Ruiz-Aceituno, L.; Sanz, M.L.; Martínez-Castro, I. Determination of free inositols and other low molecular weight carbohydrates in vegetables. *J. Agric. Food Chem.* **2011**, *59*, 2451–2455. [CrossRef] [PubMed]
14. Ruiz-Aceituno, L.; Rodríguez-Sánchez, S.; Ruiz-Matute, A.I.; Ramos, L.; Soria, A.C.; Sanz, M.L. Optimization of a biotechnologic procedure for selective fractionation of bioactive inositols in edible legume extracts. *J. Sci. Food Agric.* **2013**, *93*, 2797–280 [CrossRef] [PubMed]
15. Al-Suod, H.; Ratiu, I.-A.; Ligor, M.; Ligor, T.; Buszewski, B. Determination of sugars and cyclitols isolated from variou morphological parts of *Medicago sativa* L. *J. Sep. Sci.* **2017**, *41*, 1118–1128. [CrossRef]
16. Carrero-Carralero, C.; Mansukhani, D.; Ruiz-Matute, A.I.; Martínez-Castro, I.; Ramos, L.; Sanz, M.L. Extraction and chara terization of low molecular weight bioactive carbohydrates from mung bean (*Vigna radiata*). *Food Chem.* **2018**, *266*, 146–15 [CrossRef]
17. Mena-García, A.; Ruiz-Matute, A.I.; Soria, A.C.; Sanz, M.L. Green techniques for extraction of bioactive carbohydrate *Trends Anal. Chem.* **2019**, *119*, 115612–115622. [CrossRef]
18. Zuluaga, A.M.; Mena-García, A.; Chito-Trujillo, D.; Rada-Mendoza, M.; Sanz, M.L.; Ruiz-Matute, A.I. Development of a microwav assisted extraction method for the recovery of bioactive inositols from lettuce (*Lactuca sativa*) byproducts. *Electrophoresis* **2020**, *4* 1804–1811. [CrossRef] [PubMed]
19. Hadkar, U.; Dhruv, N.; Malode, Y.; Chavan, B. Microwave assisted extraction of phytoconstituents. *Asian J. Phytomed. Clin. Re* **2013**, *2*, 73–86.
20. Ruiz-Aceituno, L.; García-Sarrió, M.J.; Alonso-Rodriguez, B.; Ramos, L.; Sanz, M.L. Extraction of bioactive carbohydrates from artichoke (*Cynara scolymus* L.) external bracts using microwave assisted extraction and pressurized liquid extraction. *Food Chem* **2016**, *196*, 1156–1162. [CrossRef] [PubMed]
21. Zuluaga, A.M.; Mena-García, A.; Soria Monzón, A.C.; Rada-Mendoza, M.; Chito, D.M.; Ruiz-Matute, A.I.; Sanz, M.L. Microwav assisted extraction of inositols for the valorization of legume by-products. *LWT* **2020**, *133*, 109971. [CrossRef]
22. Zhang, H.-F.; Yang, X.-H.; Wang, Y. Microwave assisted extraction of secondary metabolites from plants: Current status and future directions. *Trends Food Sci Technol.* **2011**, *22*, 672–688. [CrossRef]
23. Mandal, V.; Mohan, Y.; Hemalatha, S. Microwave Assisted Extraction – An Innovative and Promising Extraction Tool for Medicin Plant Research. *Pharmacogn. Rev.* **2007**, *1*, 7–18.
24. Chemat, F.; Vian, M.A.; Fabiano-Tixier, A.S.; Nutrizio, M.; Režek Jambrak, A.; Munekata, P.E.S.; Lorenzo, J.M.; Barba, F.J.; Binello A.; Cravotto, G. A review of sustainable and intensified techniques for extraction of food and natural products. *Green Chem.* **202** *22*, 2325–2353. [CrossRef]
25. Camel, V. Microwave-assisted solvent extraction of environmental samples. *Trends Anal. Chem.* **2000**, *19*, 229–248. [CrossRef]
26. Cravotto, G.; Boffa, L.; Mantegna, S.; Perego, P.; Avogadro, M.; Cintas, P. Improved extraction of vegetable oils under high-intensit ultrasound and/or microwaves. *Ultrason. Sonochem.* **2008**, *15*, 898–902. [CrossRef]

27. Soria, A.C.; Ruiz-Aceituno, L.; Ramos, L.; Sanz, M.L. Microwave Assisted Extraction of Polysaccharides. In *Polysaccharides: Bioactivity and Biotechnology*; Mérillon, J.M., Ramawat, K.G., Eds.; Springer: Berlin/Heidelberg, Germany, 2014; pp. 987–1008, ISBN 978-3-319-03751-6.
28. Alexandru, L.; Binello, A.; Mantegna, S.; Boffa, L.; Chemat, F.; Cravotto, G. Efficient green extraction of polyphenols from post-harvested agro-industry vegetal sources in Piedmont. *C. R. Chim.* **2014**, *17*, 212–217. [CrossRef]
29. Prakash Maran, J.; Sivakumar, V.; Thirugnanasambandham, K.; Sridhar, R. Optimization of microwave assisted extraction of pectin from orange peel. *Carbohyd. Polym.* **2013**, *97*, 703–709. [CrossRef] [PubMed]
30. Eskilsson, C.S.; Björklund, E. Analytical-scale microwave-assisted extraction. *J. Chromatogr. A* **2000**, *902*, 227–250. [CrossRef]
31. Passos, C.P.; Coimbra, M.A. Microwave superheated water extraction of polysaccharides from spent coffee grounds. *Carbohydr. Polym.* **2013**, *94*, 626–633. [CrossRef] [PubMed]
32. Rodríguez-Sánchez, S.; Martín-Ortiz, A.; Carrero-Carralero, C.; Ramos, S.; Sanz, M.L.; Soria, A.C. Pressurised liquid extraction of *Aglaonema* sp. iminosugars: Bioactivity, cell viability and thermal stability. *Food Chem.* **2016**, *204*, 62–69. [CrossRef] [PubMed]
33. Jing, Y.; Lang, S.; Wang, D.; Xue, H.; Wang, X.-F. Functional characterization of galactinol synthase and raffinose synthase in desiccation tolerance acquisition in developing Arabidopsis seeds. *J. Plant Physiol.* **2018**, *230*, 109–121. [CrossRef] [PubMed]
34. Florez, N.; Conde, E.; Dominguez, H. Microwave assisted water extraction of plant compounds. *J. Chem. Technol. Biotechnol.* **2015**, *90*, 590–607. [CrossRef]
35. López-Molina, D.; Navarro-Martínez, M.D.; Rojas Melgarejo, F.; Hiner, A.N.P.; Chazarra, S.; Rodríguez-López, J.N. Molecular properties and prebiotic effect of inulin obtained from artichoke (*Cynara scolymus* L.). *Phytochemistry* **2005**, *66*, 1476–1484.
36. Castro-Puyana, M.; Marina, M.L.; Plaza, M. Water as green extraction solvent: Principles and reasons for its use. *Curr. Opin. Green Sust. Chem.* **2017**, *5*, 31–36. [CrossRef]
37. Eggleston, G.; Vercellotti, J.R. Degradation of sucrose, glucose and fructose in concentrated aqueous solutions under constant pH conditions at elevated temperature. *J. Carbohydr. Chem.* **2000**, *19*, 1305–1318. [CrossRef]
38. Möller, M.; Harnisch, F.; Schröder, U. Microwave-assisted hydrothermal degradation of fructose and glucose in subcritical water. *Biomass Bioenergy* **2012**, *39*, 389–398. [CrossRef]
39. Sanz, M.L.; del Castillo, M.D.; Corzo, N.; Olano, A. Formation of Amadori Compounds in Dehydrated Fruits. *J. Agric. Food Chem.* **2001**, *49*, 5228–5231. [CrossRef] [PubMed]
40. Villamiel, M.; Martínez-Castro, I.; Corzo, N.; Olano, A. Quantitative determination of carbohydrates in orange juice by gas chromatography. *Eur. Food Res. Technol.* **1998**, *206*, 48–51. [CrossRef]

foods

MDPI

Article

Emulsion and Surface-Active Properties of Fish Solubles Based on Direct Extraction and after Hydrolysis of Atlantic Cod and Atlantic Salmon Backbones

Silje Steinsholm [1,*], Åge Oterhals [1], Jarl Underhaug [2] and Tone Aspevik [1]

[1] Nofima AS, P.O. Box 1425 Oasen, N-5844 Bergen, Norway; aage.oterhals@nofima.no (Å.O.); tone.aspevik@nofima.no (T.A.)
[2] Department of Chemistry, University of Bergen, Allégaten 41, N-5020 Bergen, Norway; jarl.underhaug@uib.no
* Correspondence: Silje.steinsholm@nofima.no

Abstract: The focus on natural foods and "clean" labeled products is increasing and encourages development of new biobased ingredients. Fish solubles derived from downstream processing of side stream materials in the fish filleting industries have potential as emulsifiers based on their surface-active and emulsion stabilizing properties. The aim of this study was to evaluate and compare emulsion properties and critical micelle concentration (CMC) of direct protein extracts and protein hydrolysates based on fish backbones, and to identify associations between molecular weight distribution and process yield with the studied physicochemical properties. Protein extracts and enzymatic protein hydrolysates were produced based on two raw materials (cod and salmon backbones), two enzymes with different proteolytic specificity, and varying hydrolysis time. Emulsion activity index (EAI), emulsion stability index (ESI) and CMC were measured and compared with casein as a reference to protein-based emulsifiers. Protein hydrolysis was found to have negative impact on EAI and CMC, likely due to generation of small peptides disrupting the amphiphilic balance. The direct protein extracts had comparable EAI with casein, but the latter had superior ESI values. Protein hydrolysates with acceptable EAI could only be obtained at the expense of product yield. The study emphasizes the complexity of physicochemical properties of protein hydrolysates and discusses the challenges of achieving both good surface-active properties and high product yield.

Keywords: enzymatic protein hydrolysates; emulsion activity; critical micelle concentration; fish by-products; emulsion stability

1. Introduction

Emulsifiers are important ingredients in a variety of formulated food products containing two immiscible phases, such as mayonnaise, spreads, and salad dressings [1]. Their surface-activity reduces the interfacial tension between the phases and promotes the formation of stable emulsions. Present consumer attention and increasing preference for natural products and "clean" labeled food products is encouraging the development of new natural surface-active biobased ingredients [2,3]. Food-approved emulsifiers include proteins, polysaccharides, phospholipids, and synthetic surfactants [1,4]. Fish-based peptides may be a coming alternative. Given adequate surface-activity, this type of emulsifier will also add to the nutritional value while exerting a function in food formulations.

Emulsions in foods are often in the form of oil-in-water (O/W), where small droplets of lipids are distributed in a continuous aqueous phase, or water-in-oil (W/O), where oil is the continuous phase [1]. The formed emulsions are thermodynamically unstable and require the presence of an emulsifier for stabilization through reduction of surface-tension and prevention of droplet aggregation and coalescence. The amphiphilic nature of proteins facilitates adsorption at the interphase between polar and non-polar environments, after which they will reorient in a manner maximizing the contact between hydrophilic areas with the oil, while repelling other emulsion droplets [5,6]. The adsorption rate of native proteins to an interphase varies depending on the protein, but often the net charge of the proteins will not provide repulsion between droplets, causing aggregation [1]. Partial hydrolysis may improve the physicochemical properties due to exposure of hydrophobic moieties, improving the electrostatic balance, and increase the solubility and flexibility of the peptides compared to the intact protein [7,8].

Several studies have addressed the physicochemical properties of protein hydrolysates[9–21], where influences of both hydrolysis time and choice of enzyme have been assessed. The degree of hydrolysis determines the reduction in peptide molecular weight, while enzyme specificity influences the balance between hydrophilic and hydrophobic regions, both imperative for peptide surface-activity. Enzymatic protein hydrolysates of side stream products from the fish filleting industry, such as heads, backbones, and trimmings, have been proposed as a source of emulsifiers for food formulations [8]. This represents a possibility for valorization of side streams and increasing the yield of water-soluble protein, while simultaneously improving the functionality of the native proteins [22]. Hydrolysis of herring protein has been shown to improve emulsion activity and stability [23], and sardine protein hydrolysates showed better emulsion properties compared with sodium caseinate [24]. However, hydrolysis of the common emulsifiers casein and whey have not always been found to improve physicochemical properties (emulsion capacity and stability), having emulsifying abilities either inferior or comparable to that of the native proteins [24]. In general, large peptides (>2–4 kDa) are essential for proper functionality, and it has been suggested that peptides should contain more than 20 amino acids to exhibit good emulsifying capability [16]. In addition, peptide–peptide interactions are particularly important [5,10,16].

Several approaches can be used in the determination of surface-activity and emulsion properties of proteins and peptides. Emulsion activity (emulsion activity index; EAI) determines the obtainable interfacial area between oil and water per unit weight of protein or product, and emulsion stabilizing ability (emulsion stability index; ESI) indicates the emulsifying effect over time [25]. Critical micelle concentration (CMC) indicates the minimum concentration of a product needed for maximum reduction of the surface-tension and can be assessed by different methods based on fluorescence, conductivity, surface-tension, or ^{1}H nuclear magnetic resonance (NMR) [26,27]. High CMC values imply poor surface-activity, i.e., a high concentration of the given surfactant is needed to reduce the surface-tension [10,27]. Measurements of properties related to emulsion capabilities of protein hydrolysates are challenging to standardize. Pearce and Kinsella [25] showed that EAI results are dependent on assay variations, such as homogenization factors and protein concentration, and the latter was confirmed by Nalinanon et al. [17]. This makes interstudy comparisons difficult and may add to the contradictory results from previous emulsion studies of protein hydrolysates.

There is a lack of knowledge on the effect of raw material and process variables on properties related to surface-activity of fish solubles. The aims of this study were (1) to evaluate and compare physicochemical properties (i.e., emulsion activity (EAI), stability index (ESI) and CMC) based on direct extraction and enzymatic hydrolysis of salmon and cod backbones, and (2) to assess the association between peptide molecular weight distribution, physicochemical properties, and process yields.

2. Materials and Methods

2.1. Materials

Atlantic salmon (*Salmo salar*) and cod (*Gadus morhua*) backbones were kindly provided by Sotra Fiskeindustri AS (Glesvær, Norway) and Halstensen Granit AS (Bekkjarvik Norway), respectively. The raw materials were milled on a Comitrol 1700 (Urschel laboratories, Chesterton, IN, USA), vacuum packed, and stored at −20 °C until use. The applied enzymes were Bromelain BR1200 (EC 3.4.22.32, Enzybel, Waterloo, Belgium) and FoodPro PNL (EC 3.4.24.28, DuPont, Wilmington, DE, USA). Refined rapeseed oil was purchased at a local supermarket (Rema 1000 store brand, Kjerreidviken, Norway). Peptide standards were purchased from Sigma-Aldrich (Oslo, Norway) except lysozyme (Fluka biochemicals, Buchs, Switzerland) and Alberta standards (Alberta Peptide Institute, Department of Biochemistry, University of Alberta, Edmonton, AB, Canada). Technical grade Tween 20 (VWR, Oslo, Norway) and bovine casein (Sigma, Oslo, Norway) were applied in the emulsion assay. All other chemicals were analytical or food grade.

2.2. Chemical Analyses

Analysis of nitrogen (N) was performed by the Kjeldahl method [28] and the crude protein level determined based on substrate specific N-to-protein conversion factor [29]. Amino acid composition was quantified by High performance liquid chromatography (HPLC) using fluorescence detection with excitation/emission at 250/395 nm. Proteins were hydrolyzed to free amino acids with 6N HCl and amino acids derivatized with 6-aminoquinolyl-*N*-hydroxysuccinimidyl carbamate before passing through the HPLC column (Waters Accq Tag 3.9 × 150 mm, Milford, MA, USA) and detector [30]. Asparagine and glutamine were estimated based on the release of ammonia in the HCl digest compared to a neutral control sample [31]. Released ammonia was quantified by the method of Conway and Byrne [32]. Analysis of molecular weight distribution (MWD) was performed by HPLC (1260 series HPLC, Agilent Technologies, Santa Clara, CA, USA) using size exclusion chromatography [33], as described by Oterhals and Samulesen [34]. All chemical analyses were performed in duplicate with predetermined allowances for replicate variation.

2.3. Enzymatic Protein Hydrolysis

Raw materials were thawed overnight at 4 °C. The raw material was mixed with purified water (1:1) and transferred to a Distek Model 2500 Dissolution System (Distek Inc., North Brunswick, NJ, USA). The slurry was heated to 50 °C at continuous stirring (70 rpm) before adding 10 U enzyme per gram protein [35]. The proteolytic reactions were terminated after 5, 10, 30 or 60 min by heating to >90 °C in a microwave oven (Menumaster commercial, Cedar Rapids, IA, USA) for a minimum 10 min. The slurry was cooled to <40 °C in a water bath before phase separation by centrifugation at 15,000× *g* for 20 min (Sorvall, LYNX 6000, Thermo Scientific, Waltham, MA, USA). Direct protein extracts by thermal coagulation were produced with the same method, with the exception of enzyme addition, of both raw materials. The water phase was filtered through a Seitz-T2600 filter (Mall Corporation, East Hills, NY, USA) to remove larger particles and thereafter subjected to 0.1 μm cross flow membrane filtration (Centramate 500S Tangential Flow Filtration System, Pall, Port Washington, NY, USA) to remove remnant fine particles and fat. The final hydrolysates were stored at −20°C until further use.

Protein recovery (PR) was determined based on protein content in the filtered hydrolysate compared to that in the raw material:

$$\text{PR} = \frac{\text{Protein in the hydrolysate} \times \text{g hydrolysate}}{\text{Protein in the raw material} \times \text{g raw material}} \times 100\% \tag{1}$$

volume of collected hydrolysate after membrane filtration was corrected for the remaining dead volume in the Centramate filter apparatus.

2.4. Determination of Critical Micelle Concentration by NMR Spectroscopy

All samples were diluted 3:4 with 400 mM sodium phosphate buffer (pH 6.5) containing 15% D_2O and 2,2-dimethyl-2-silapentane-5-sulfonate (DSS). The pH was adjusted to 6.5 with 0.1 M HCl. Dilution series of 10 samples were prepared from all hydrolysates with the final concentration of 100 mM sodium phosphate buffer and 7.5% D_2O. A volume of 600 µL was transferred to 5 mm NMR tubes. ^{1}H spectra were acquired at 300 K with a Bruker AVANCE NEO ultrashielded 600 MHz spectrometer with cryoprobe using Bruker pulse program zgesgppe (Karlsruhe, Germany). Acquisition parameters were set to four dummy scans, 32 real scans, 1 s relaxation delay, 32k time-domain points, and spectral width of 11.9 ppm. The NMR spectra were processed using TopSpin (v. 4.0.4, Bruker BioSpin, Karlsruhe, Germany). Exponential line broadening of 0.5 Hz was applied prior to Fourier transformation, and the chemical shifts were referenced to DSS.

The CMC was estimated by plotting the ^{1}H shift of the lactate methyl group as a function of log protein concentration [10]. The methyl resonance of lactate was visible around 1.3 ppm for all samples. Separate linear trend lines were drawn for the lag phase and the exponential phase in the plot. The data points used for the lag phase tangent were adjusted to obtain best fit by removing points in the transition between lag and exponential phase. The intercept of these lines indicated the start of protein aggregation and thus the critical micelle concentration. Propagation of uncertainty assuming independent variables indicated standard measurement error $< \pm 0.2$ based on the fitted regression lines and standard error for the coefficient provided by the LINEST function in Microsoft Excel (v. 2013). Aspevik et al. [10] estimated the standard deviation for the used protocol to be 0.5 g/L.

2.5. Emulsion Properties

Emulsion activity index (EAI) and emulsion stability index (ESI) were determined based on the method by Pearce and Kinsella [25] and modified by Liceaga-Gesualdo and Li-Chan [23]. The hydrolysates were diluted to 1.0% protein in purified water and pH adjusted to 6.5 with 0.1 M HCl. Tween 20 was used as control to ensure method repeatability, and casein (1% solubilized in 50 mM potassium phosphate) as a reference to commercial emulsifiers. Emulsions were made by adding 2 mL of rapeseed oil to 6 mL of standardized hydrolysate and homogenized (T 10 basic Ultra-Turrax, Ika, Staufen, Germany) at 16,000 rpm for 1 min in a glass container. Emulsions were made in triplicate. Immediately after homogenization, 25 µl was collected from the bottom of the mixture, added to 5 mL 0.1% sodium dodecyl sulfate (SDS) and the mixture inverted two times before measurement of EAI. Sample collection was repeated after 10 and 30 min for determination of ESI. Absorbance was measured at 500 nm and EAI was calculated based on the following equation:

$$\text{EAI}\left(\text{m}^2(\text{g protein})^{-1}\right) = \frac{2 \times \frac{2.303\text{A}}{\text{l}} \times \text{df}}{\varphi \times \text{c}} \tag{2}$$

where A = absorbance measured at 0, 10 or 30 min; l = path length of cuvette in m; df = dilution factor (200); φ = oil phase volume of total mixture volume; c = weight of protein per unit volume of aqueous sample before emulsion. ESI was calculated based on the percent EAI remaining after a defined stagnant standing period:

$$\text{ESI}(\%) = 100 - \left(\frac{\text{EAI} - \text{EAI}_t}{\text{EAI}} \times 100\right) \tag{3}$$

where EAI_t = EAI value of samples collected after 10 or 30 min.

2.6. Statistics

Analysis of variance (ANOVA) was performed using Minitab (v. 19.2, Pennsylvania State University, PA, USA). One-way ANOVA was used to determine significant product differences of EAI and ESI results. Two-way ANOVA determined significant parameters. Hydrolysates were used to model effect of raw material, enzyme, and hydrolysis time. Tukey's pairwise comparison was used when significance ($p < 0.05$) was found.

Principal component analysis (PCA) was used to evaluate relationships between EAI, ESI, and MWD using Unscrambler (v10.4.1, Camo, Oslo, Norway). All data were unit variance scaled and centered before analysis.

3. Results and Discussion

3.1. Raw Material and Hydrolysate Composition

Salmon and cod backbones showed similar amino acid compositions (Table 1), with slightly higher levels in the cod raw material, reflecting higher protein concentration in the latter substrate. Both substrates contained high levels of glycine, proline, and hydroxyproline, ascribed to the high proportion of bones and connective tissue protein. N-to-protein conversion factors (f_N) for cod and salmon backbones were calculated to 5.5 and 5.2, respectively, in agreement with previous findings for pure muscle proteins from the two species [35], and illustrated the deviance of fish raw material from the commonly used factor of 6.25. The use of substrate specific conversion factors facilitated a more accurate quantification of protein content, and thus enzyme addition on protein basis and standardization of hydrolysis conditions in studies applying different protein sources [30].

Table 1. Amino acid composition in the raw material (g kg^{-1}; n = 2) and substrate specific nitrogen-to-protein conversion factor data of salmon (*Salmo salar*) and Cod (*Gadus morhua*) backbones.

AA (g/kg) **	Cod Backbone	Salmon Backbone
Alanine	11.0	9.9
Arginine	11.0	9.0
Asparagine *	7.1	6.3
Aspartate	8.9	7.7
Glutamate	16.1	12.0
Glutamine *	7.9	7.0
Glycine	17.0	14.0
Histidine	3.3	3.6
Isoleucine	6.4	5.7
Leucine	12.0	9.6
Lysine	13.0	11.0
Methionine	5.4	4.6
Phenylalanine	5.9	5.2
Proline	8.8	7.8
Serine	8.6	6.5
Threonine	6.9	6.2
Tyrosine	6.9	4.0
Valine	7.4	6.7
Hydroxyproline	3.8	3.5
Nitrogen	27	25
NH3 (acid digest)	1.8	1.6
Total AA	167.4	140.3
Total *n*	27.0	25.0
f_N ***	5.5	5.2

* Calculated based on released NH_3, and assuming 1:1 ratio of released NH_3 between Asp and Glu [29]. ** Amino acid. *** Nitrogen-to-protein conversion factor.

The enzymes were added based on similar activity-to-protein ratio [31], and the resulting peptide molecular weight distribution (MWD; Table 2) showed an increase in smaller peptides as the hydrolysis progressed. Furthermore, levels of molecules <0.2 kDa were higher in products based on FoodPro PNL compared to the equivalent Bromelain hydrolysate, indicating some exopeptidase activity in the former enzyme [31]. In general, hydrolysates based on cod backbone contained a larger proportion of peptides, >1 kDa, compared with salmon. For both raw materials, the PR increased with prolonged hydrolysis time, with Bromelain giving slightly higher levels compared with FoodPro PNL (Table 3). This is possibly explained by the broad specificity of Bromelain and efficiency on connective tissue proteins [36]. As expected, the direct protein extracts contained mostly large peptides, >10 kDa, and small molecules, <0.2 kDa, characteristic for a product based on direct thermal coagulation and separation [10,34].

able 2. Apparent molecular weight distribution (MWD; kDa) and average amino acid units in the molecular size of direct protein xtracts (-ext) and hydrolysates made from cod (C) and salmon (S) backbones with FoodPro PNL (F) and Bromelain (B) for 5, 10, 30 nd 60 min.

MWD (%)	C-ext	CF05	CF10	CF30	CF60	CB05	CB10	CB30	CB60	S-ext	SF05	SF10	SF30	SF60	SB05	SB10	SB30	SB60	AA *
>20	8.9	1.3	1.2	0.4	0.2	1.3	1.4	0.8	0.3	4.9	2.6	1.5	0.4	0.1	1.1	0.8	0.1	<0.1	>198
15–20	3.9	1.1	0.9	0.3	0.1	1.0	0.9	0.5	0.2	2.2	1.7	1.1	0.4	0.1	0.6	0.4	0.1	<0.1	138
10–15	5.7	2.9	2.7	1.1	0.5	3.0	2.6	1.3	0.5	3.1	4.1	3.1	1.3	0.4	1.2	0.8	0.2	<0.1	99
8–10	3.4	3.0	2.9	1.5	0.8	3.5	3.2	1.8	0.8	1.7	4.0	3.4	1.9	0.9	1.0	0.7	0.2	0.1	71
6–8	4.0	6.0	6.1	3.9	2.5	7.8	7.4	5.1	2.8	2.2	7.0	6.5	4.5	2.7	2.5	2.0	0.6	0.3	55
4–6	3.8	12.0	12.6	9.7	7.2	15.9	16.0	13.7	10.0	2.4	11.6	11.2	8.9	6.7	7.6	6.6	2.9	1.7	32
2–4	3.7	21.4	23.0	22.8	21.2	26.4	27.9	30.0	29.3	2.5	18.6	20.0	20.0	18.2	23.3	23.0	16.0	12.0	20
1–2	1.9	16.1	17.2	20.2	21.4	14.6	16.2	20.7	25.0	1.3	12.3	14.0	16.9	18.4	22.2	23.7	25.7	24.5	12
0.5–1	1.2	10.0	11.0	14.9	17.5	6.2	6.9	9.7	13.3	0.8	7.7	9.2	13.0	16.0	12.8	14.3	21.1	24.2	5.9
0.2–0.5	3.3	6.6	7.1	10.3	13.0	3.3	3.4	4.4	6.0	18.4	10.3	11.0	13.6	16.4	10.7	11.5	16.5	20.0	2.8
<0.2	60.3	19.7	15.4	14.8	15.7	17.0	14.1	12.1	11.9	60.7	20.2	19.1	19.1	20.1	16.9	16.4	16.4	17.2	1.0

* Estimated average number of amino acid units in the molecular size group based on weighted average MW [30].

Table 3. Measured emulsion activity index (EAI *) and stability index after 10 and 30 min (ESI-10/30 *), critical micelle concentration (CMC **) and protein recovery (PR) of hydrolysates and direct protein extracts (-ext) based on cod (C) and salmon (S) backbones with FoodPro PNL (F) and Bromelain (B) for 5, 10, 30 and 60 min of hydrolysis.

	EAI (m^2/g)	ESI-10	ESI-30	CMC (g/l)	PR (%)
C-ext	13 ± 0.9^{a}	21 ± 1.9^{cde}	16 ± 3.0^{de}	1.6	6.2
CF05	11 ± 1.2^{bcd}	33 ± 2.7^{a}	25 ± 0.6^{ab}	3.7	18.1
CF10	9.0 ± 0.2^{cd}	18 ± 1.9^{def}	9 ± 0.6^{efg}	4.8	23.4
CF30	11 ± 0.7^{bcd}	14 ± 2.1^{ef}	7 ± 1.3^{fgh}	6.0	31.4
CF60	9.0 ± 0.2^{d}	27 ± 0.5^{abcd}	22 ± 2.9^{abc}	6.2	35.6
CB05	11 ± 0.5^{bcd}	31 ± 1.7^{abc}	22 ± 2.9^{bcd}	5.1	24.0
CB10	11 ± 0.7^{bcd}	25 ± 1.6^{abc}	17 ± 1.5^{cde}	5.4	27.0
CB30	11 ± 0.4^{bcd}	33 ± 1.4^{abc}	26 ± 3.1^{ab}	6.0	31.5
CB60	11 ± 0.1^{abc}	31 ± 2.2^{a}	27 ± 1.6^{a}	6.6	36.6
S-ext	12 ± 1.2^{ab}	10 ± 1.0^{f}	6 ± 1.8^{gh}	1.8	7.0
SF05	10 ± 1.2^{bcd}	11 ± 2.4^{f}	6 ± 1.0^{gh}	5.8	26.5
SF10	10 ± 0.3^{bcd}	12 ± 0.0^{f}	6 ± 1.2^{gh}	5.3	25.8
SF30	10 ± 0.4^{bcd}	30 ± 2.2^{ab}	23 ± 1.6^{abc}	6.7	31.1
SF60	10 ± 0.3^{bcd}	32 ± 3.6^{a}	25 ± 3.7^{ab}	6.8	33.7
SB05	11 ± 0.3^{bcd}	11 ± 3.5^{f}	4 ± 0.7^{h}	5.3	26.1
SB10	10 ± 0.5^{bcd}	13 ± 0.6^{ef}	5 ± 1.0^{h}	6.2	30.0
SB30	10 ± 0.5^{bcd}	22 ± 1.7^{bcd}	13 ± 0.4^{ef}	7.6	40.0
SB60	11 ± 0.5^{bcd}	31 ± 3.3^{ab}	22 ± 1.4^{abc}	7.6	40.8

* Different letters indicate statistically different values ($p \leq 0.05$) by one-way ANOVA and Tukey's pairwise comparison. ** Measured based on single samples. Standard deviation was estimated to 0.5 g/L by Aspevik et al. [10].

All products were purified by microfiltration to eliminate suspended solids and residual lipids with possible bias effects in the surface-activity assays. The protein recoveries (Table 3) were lower than observed in earlier studies without a microfiltration step [10], particularly for the direct extracts (PR = 6–7% compared to expected approximate 20%). This confirms a partial retention of large proteins and peptide fragments by microfiltration, as earlier reported [37]. The shift in MWD toward smaller peptides probably influenced the surface-active properties of the respective products; however, effects of membrane filtration were outside the scope of this study.

3.2. Associations between MWD and Physicochemical Properties

Principal component analysis (PCA; Figure 1) was used to evaluate associations between EAI, ESI, CMC, PR, and MWD of the hydrolysates. Data from the direct protein extracts were excluded due to the deviant MWD and PR compared with the hydrolysates (Tables 2 and 3), adding too much leverage to the model and dominating the variation of the two variables. Based on the score plot, hydrolysates with similar or different properties could be identified. Two principal components (PCs) were found to be relevant for the interpretation of results. The first and second PCs explained 58% and 21%, respectively.

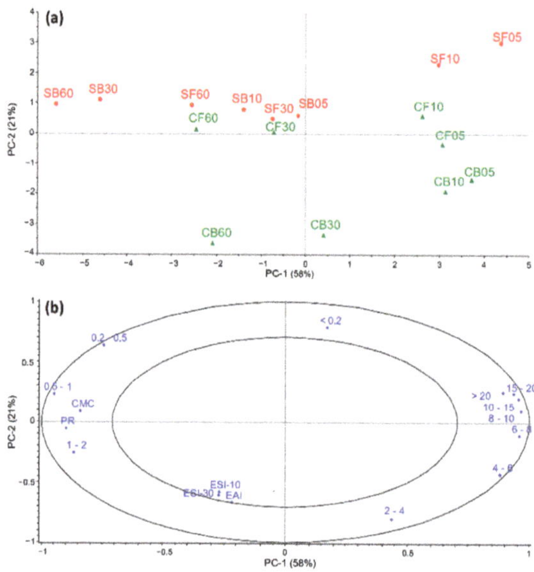

Figure 1. Principal component analysis score plot (**a**) shows similarities and differences between salmon (S) and cod (C) backbone hydrolysates made with FoodPro PNL (F) and Bromelain (B) for 5, 10, 30 and 60 min. The correlation loading plot (**b**) illustrates associations between molecular weight distribution (kDa), protein recovery (PR), critical micelle concentration (CMC), emulsion stability after 10 min (ESI-10), emulsion stability after 30 min (ESI-30), and emulsion activity index (EAI). The two ellipses represent 50% and 100% of explained variance.

In the score plot (Figure 1a), PC-1 mainly explains the effect of hydrolysis time, while the raw material variation is explained by PC-2. The correlation loading plot (Figure 1b) mostly shows a product separation based on CMC, PR, and MWD. High values for the two former variables and small molecules of 0.5–2 kDa were associated and negatively correlated with molecules of 4–>20 kDa, in agreement with Aspevik et al. [10]. The emulsion responses were less than 50% explained by the model, thus interpretation should be done with care. The loading plot shows no positive correlations between emulsion properties and specific MW groups. This was also the case in studies on salmon muscle protein hydrolysates [15], and whey and casein protein hydrolysates [24]. The results indicate that extended hydrolysis is detrimental to surface-activity, expressed by CMC. The MWD (Table 2) and considerably lower CMC values (Table 3) for the direct protein extracts supported this conclusion.

3.3. Effect of Process Parameters on Emulsion Properties

The EAI of the products (Table 3) showed small differences between the enzymatic hydrolysates, but the direct protein extracts gave the highest values for both cod and salmon. This may be attributed to the higher levels of large peptides (>10 kDa) being sufficiently flexible for effective interfacial surface coverage. Furthermore, a decrease in surface hydrophobicity of the hydrolysates may also add to this observation, as discussed for whey hydrolysates [20]. Negligible differences between the EAI (Table 3) of the hydrolysates were observed, with no clear pattern of differences influenced by raw material, enzyme, or hydrolysis time (Figure 2).

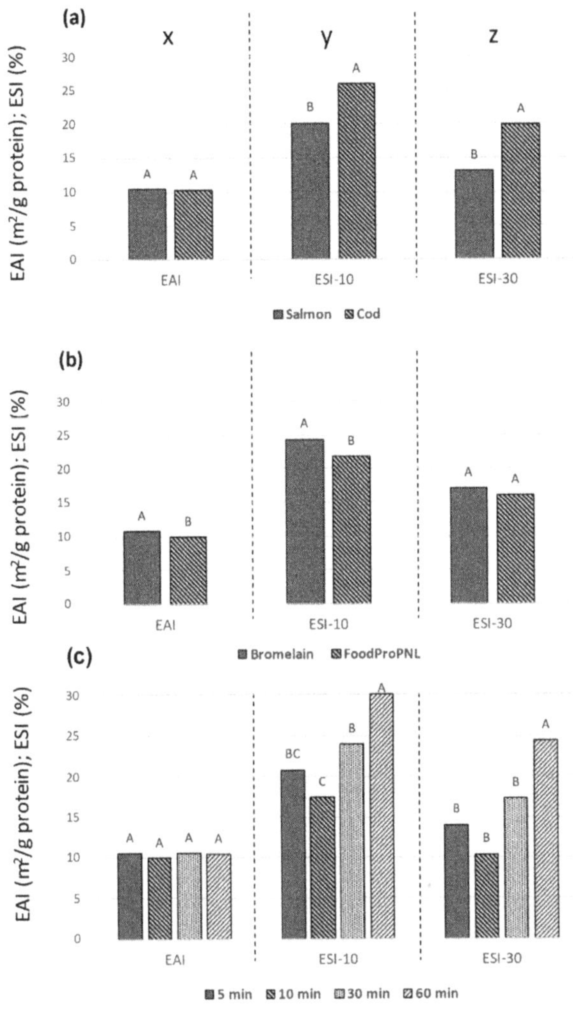

Figure 2. Mean emulsion activity index (EAI) and emulsion stability index (ESI) after 10 and 30 min for the raw materials salmon and cod (**a**), the enzymes FoodPro PNL and Bromelain (**b**), and hydrolysis times (**c**). Different letters indicate statistical effect of the hydrolysis parameter on the emulsion variable. EAI (x), EAI-10 (y) and EAI-30 (z) are separate statistical entities, indicated by the dotted lines.

Casein is an excellent emulsifier in milk-based products [6], and the EAI of casein was measured at 16 ± 1 m^2/g protein, only slightly higher than the direct protein extracts in this study (EAI = 12–13, Table 3). We have found few studies comparing the EAI of fish-based protein hydrolysates with direct protein extraction of the raw material. Contrary to this study, Liceaga-Gesualdo and Li-Chan [23] showed enhanced EAI of herring hydrolysates compared to this type of reference sample. More common is to use a commercial protein emulsifier as reference. Tan et al. [21] found that restricted hydrolysis on catfish gave EAI and ESI comparable to those of soy protein isolate. However, Alves et al. [9] found the EAI of soy proteins to be superior to that of chicken blood hydrolysates and associated this to large interfacial areas of the soy proteins. The interfacial properties of proteins could possibly explain the relatively good results for the casein protein and the direct thermal extracts compared to the hydrolysates in this study. The mentioned studies [6,9,21,23] had assay variations comparable to the current work, such as sample dilution media, protein concentration, and pH, likely influencing the results.

The ESI showed more variation between the samples compared to the EAI (Table 3), with lowest values obtained by the direct protein extracts. The low stability of the two latter emulsions may be due to a combination of electrostatic attraction between unfolded protein and interactions with other small molecules present in the samples [1]. An increase in ESI with prolonged hydrolysis time for the salmon hydrolysates was observed, suggesting that a higher release of small molecules is required for stabilization, compared with shorter hydrolysis time (Figure 2). Other studies have reported a decrease in emulsion stability in hydrolysates based on salmon heads [13], muscle proteins [15], and chicken blood [9] when the degree of hydrolysis is increased. They attributed the reduction in ESI to a reduction in interfacial tension and increased hydrolysis, indicating a loss of emulsion stabilizing properties when the salmon peptides are substantially hydrolyzed.

The ESI values for the cod hydrolysates, on the other hand, were more ambiguous, with low levels at intermediate hydrolysis times. The highest observed levels were, however, similar for both salmon and cod, and the values decreased after 30 min holding time. The ESI values of casein were about two times higher compared with the hydrolysates, with values for ESI-10 = 54 $\pm$ 5 and ESI-30 = 40 $\pm$ 4, likely due to an appropriate balance of hydrophobic regions [1] and high film viscosity.

ANOVA of the individual enzymatic hydrolysis parameters demonstrated that there were no significant effects of species or hydrolysis time on EAI (Figure 2(ax,cx)), whereas Bromelain gave significantly higher values compared with FoodPro PNL (Figure 2(bx)). The proteases Bromelain and FoodPro PNL (formerly named Protex 7L) were chosen based on studies suggesting good emulsion properties and CMC values in hydrolysates based on these enzymes [10,12]. The observed difference, although small, was in agreement with a study on tilapia hydrolysates [12], where Bromelain gave superior emulsion properties of the four proteases tested.

All hydrolysis parameters significantly influenced the ESI-10 levels (Figure 2(ay–cy)). where cod, Bromelain, and 60 min hydrolysis were superior to salmon, FoodPro PNL, and shorter hydrolysis time, respectively. This suggests that smaller peptides may be important for emulsion stability, in agreement with other studies on fish-based substrates [17,18]. On the other hand, there have been studies suggesting a general decrease in emulsion properties of fish-based substrates as the hydrolysis progressed [14,15], but very high protease concentration [14] and different emulsion assays [15] were applied. The PCA-plot (Figure 1) indicates a negative association between MW < 0.2 kD and ESI. Characteristic for the two direct protein extracts is a very high content (60%) of this MW fraction (Table 2) and a low ESI. Combined, this indicates a negative effect of free amino acids on EAI and a positive effect of higher molecular weight peptides in general. No correlations were found between specific peptide fractions and ESI; however, the effect might reflect a low contribution to film viscosity of free amino acids compared to peptides.

Interpretation and comparison between separate studies should be made with caution. The protein concentration applied in EAI assays strongly influences the results [17], along with pH [18] and the equipment used [25]. These factors have been taken into consideration when assessing similarities with other studies, but the challenges emphasize the need for more standardized assay methodology. The process of emulsion formation and stabilization is complex with many influencing factors [5], especially peptide–peptide interactions [16]. Furthermore, the relatively low ESI of the direct extracts (Table 3) shows that the formation and stability of emulsions cannot be seen as co-dependent responses, as they appear to depend on different peptide properties, in agreement with previous studies [24,38].

3.4. Critical Micelle Concentration of Fish Protein Hydrolysates

The use of ^{1}H NMR is a well-established method for determination of CMC [10,27]. In this study, the chemical shift of lactate was used to measure changes in the chemical environment due to micelle formation (Figure 3). A full overview of the chemical shifts with the corresponding protein concentration can be found in Table S1. A low CMC is favorable and implies that less of the surfactant is needed to obtain maximum reduction of the surface-tension. The lowest value of CMC was observed for the direct extracts (Table S1) and indicated that the undigested proteins present were flexible enough to exert a better reduction of surface-tension than their peptide moieties. The NMR technology measures a change in the chemical environment; however, it cannot discern if the micelles or aggregates are homogenously distributed, which indicates electrostatic repulsion necessary for emulsifying capabilities. This may add to an explanation of the lack in correlation between CMC and ESI (Figure 1).

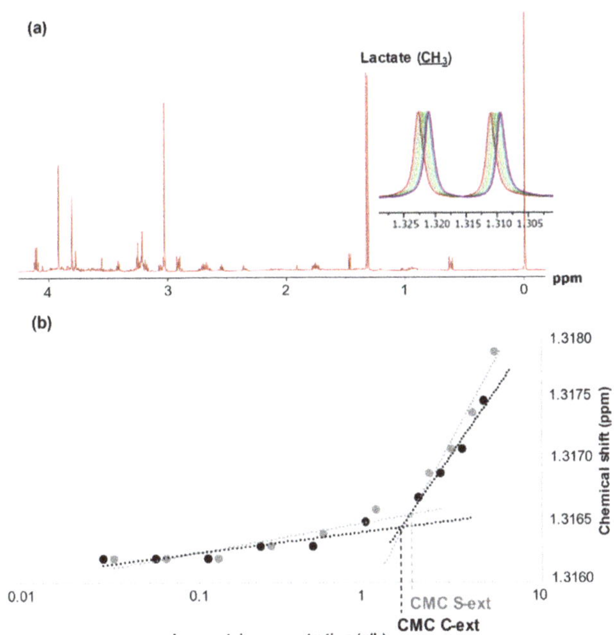

Figure 3. (**a**) ^{1}H NMR spectra showing the methyl resonance of lactate around 1.3 ppm, with an insert illustrating the change in chemical shift of lactate with decreasing protein concentration of the salmon protein extract dilution series. (**b**) Determination of the critical micelle concentration (CMC) of the direct protein extracts based on cod (C-ext) and salmon (S-ext).

An increase in hydrolysis time, and thus reduction in peptide size, gave higher CMC values (Table 3). This was in agreement with a previous observation [10] and likely due to a decrease in the amphiphilic nature of the peptides by extended hydrolysis [6]. The CMC of cod hydrolysates was slightly lower compared with the corresponding salmon hydrolysates and may be explained by the generally higher content of larger peptides in cod-based hydrolysates (Table 2). Furthermore, the CMC values of hydrolysates based on Bromelain were higher than the corresponding FoodPro PNL hydrolysates, probably due to the broad specificity of the former enzyme, leading to more disruption of the hydrophobic areas in the peptides [36]. The obtained CMC values were lower than those reported by Aspevik et al. [10] for salmon heads and backbones, which ranged from 6 g/L in the control sample (without proteolysis) to 11.5 g/L after prolonged hydrolysis with FoodPro PNL. The respective values of the current study are 1.8 and 6.8 g/L. A major difference between the two studies was the use of a 100 kDa [10] and 0.1 μm (this study) membrane filter to remove residual lipids and fine particles in the hydrolysate before measurement of physicochemical properties. A microfiltration step is needed to remove interfering compounds; however, it will also partly remove higher molecular weight peptides with negative impact on CMC. The variation in CMC between the two studies shows that a 100 kDa filter removes more of the high MW molecules required for low CMC values.

The negative correlation between PR and low CMC (Figure 1) suggests that a compromise must be met between high surface-activity and product yield. Although by restricting hydrolysis to a degree where peptide surface-activity is retained, the lower PR may be compensated by introduction of a cascade approach where peptide fractions with different physicochemical properties are obtained after successive hydrolysis steps to improve the overall process yield. Additional studies on droplet size distribution and reduction in interfacial tension may be included to further elucidate the physicochemical properties.

4. Conclusions

No associations between CMC and emulsion properties (EAI and ESI) of protein hydrolysates based on cod and salmon backbones were observed. Low CMC, implying good surface-activity, was correlated with peptides > 4kDa and hydrolysates of restricted proteolysis. The lowest CMC and highest EAI values were obtained for products based on direct protein extraction without hydrolysis and reflected a negative effect of hydrolysis on CMC and EAI. The ESI values of the hydrolysates were both increased and reduced compared with direct protein extraction. The process combination of cod, Bromelain, and 60 min of hydrolysis was superior to salmon, Food Pro, and shorter hydrolysis times with respect to this property. The EAI values of direct protein extracts were slightly lower than casein; however, ESI values were less competitive. Hydrolysates showed both lower EAI and ESI values compared to casein. Direct protein extraction gives superior physicochemical properties measured as CMC and EAI; however, it also results in lower ESI and product yield compared with hydrolysis. This is further reinforced by microfiltration to remove residual lipids and fine particles. A cascade approach is suggested as a potential method both to improve product yield and optimize emulsifier properties.

Supplementary Materials: The following are available online at https://www.mdpi.com/article/10.3390/foods10010038/s1, Table S1: Chemical shifts of lactate methyl resonance.

Author Contributions: Conceptualization, S.S., Å.O. and T.A.; Formal analysis, S.S.; Investigation, S.S.; Methodology, S.S., Å.O., J.U. and T.A.; Project administration, T.A.; Supervision, Å.O., J.U. and T.A.; Visualization, S.S.; Writing—original draft, S.S.; Writing—review and editing, S.S., Å.O., J.U. and T.A. All authors have read and agreed to the published version of the manuscript.

Funding: The work is funded by Nofima through the project "PepTek," and The Research Council of Norway through the project WHITEFISH (project number 280498) is greatly acknowledged. The Research Council of Norway is acknowledged for the support to the Aquafeed Technology Centre (ATC; project number 245883/F50) and the Norwegian NMR Platform (NNP; project number 226244/F50). Bergen Research Foundation (project code BFS-NMR-1) and Sparebankstiftinga Sogn og Fjordane (project code 509–42/16) are acknowledged for their funding of the 600 MHz NMR Spectrometer.

Data Availability Statement: The data presented in this study are available in the article and the supplementary material.

Acknowledgments: The authors want to thank the laboratory personnel at Nofima BioLab for their skillful and accurate work on the chemical analyses. We would also like to thank Eva Veiseth-Kent at Nofima Ås for help with the emulsion assay.

Conflicts of Interest: The authors declare no conflict of interest.

References

1. McClements, D.J. *Food Emulsions—Principles, Practices and Techniques*, 3rd ed.; CRC Press: Boca Raton, FL, USA, 2016.
2. Ozturk, B.; McClements, D.J. Progress in natural emulsifiers for utilization in food emulsions. *Curr. Opin. Food Sci.* **2016**, *7*, 1–6.
3. Berton-Carabin, C.; Schroen, K. Towards new food emulsions: Designing the interface and beyond. *Curr. Opin. Food Sci.* **2019**, *27*, 74–81.
4. Kralova, I.; Sjoblom, J. Surfactants Used in Food Industry: A Review. *J. Disper. Sci. Technol.* **2009**, *30*, 1363–1383.
5. Dexter, A.F.; Middelberg, A.P.J. Peptides as functional surfactants. *Ind. Eng. Chem. Res.* **2008**, *47*, 6391–6398.
6. McClements, D.J. Protein-stabilized emulsions. *Curr. Opin. Colloid Interface Sci.* **2004**, *9*, 305–313.
7. Panyam, D.; Kilara, A. Enhancing the functionality of food proteins by enzymatic modification. *Trends Food Sci. Technol.* **1996**, *7*, 120–125.
8. Halim, N.R.A.; Yusof, H.M.; Sarbon, N.M. Functional and bioactive properties of fish protein hydolysates and peptides: A comprehensive review. *Trends Food Sci. Technol.* **2016**, *51*, 24–33.
9. Da Silva Bambirra Alves, F.E.; Carpiné, D.; Teixeira, G.L.; Goedert, A.C.; de Paula Scheer, A.; Ribani, R.H. Valorization of an Abundant Slaughterhouse By-product as a Source of Highly Technofunctional and Antioxidant Protein Hydrolysates. *Waste Biomass Valorization* **2020**. [CrossRef]
10. Aspevik, T.; Totland, C.; Lea, P.; Oterhals, A. Sensory and surface-active properties of protein hydrolysates based on Atlantic salmon (Salmo salar) by-products. *Process Biochem.* **2016**, *51*, 1006–1014.
11. Chobert, J.M.; Bertrandharb, C.; Nicolas, M.G. Solubility and Emulsifying Properties of Caseins and Whey Proteins Modified Enzymatically by Trypsin. *J. Agric. Food Chem.* **1988**, *36*, 883–892.
12. Elavarasan, K.; Kumar, V.N.; Shamasundar, B.A. Antioxidant and Functional Properties of Fish Protein Hydrolysates from Fresh Water Carp (Catla catla) as Influenced by the Nature of Enzyme. *J. Food Process. Pres.* **2014**, *38*, 1207–1214.
13. Gbogouri, G.A.; Linder, M.; Fanni, J.; Parmentier, M. Influence of Hydrolysis Degree on the Functional Properties of Salmon Byproducts Hydrolysates. *J. Food Sci.* **2004**, *69*, C615–C622.
14. Klompong, V.; Benjakul, S.; Kantachote, D.; Shahidi, F. Antioxidative activity and functional properties of protein hydrolysate of yellow stripe trevally (Selaroides leptolepis) as influenced by the degree of hydrolysis and enzyme type. *Food Chem.* **2007**, *102*, 1317–1327.
15. Kristinsson, H.G.; Rasco, B.A. Biochemical and functional properties of Atlantic salmon (Salmo salar) muscle proteins hydrolyzed with various alkaline proteases. *J. Agric. Food Chem.* **2000**, *48*, 657–666. [PubMed]
16. Lee, S.W.; Shimizu, M.; Kaminogawa, S.; Yamauchi, K. Emulsifying Properties of Peptides Obtained from the Hydrolyzates of Beta-Casein. *Agric. Biol. Chem.* **1987**, *51*, 161–166.
17. Nalinanon, S.; Benjakul, S.; Kishimura, H.; Shahidi, F. Functionalities and antioxidant properties of protein hydrolysates from the muscle of ornate threadfin bream treated with pepsin from skipjack tuna. *Food Chem.* **2011**, *124*, 1354–1362.
18. Pacheco-Aguilar, R.; Mazorra-Manzano, M.A.; Ramirez-Suarez, J.C. Functional properties of fish protein hydrolysates from Pacific whiting (Merluccius productus) muscle produced by a commercial protease. *Food Chem.* **2008**, *109*, 782–789.
19. Quaglia, G.B.; Orban, E. Influence of Enzymatic-Hydrolysis on Structure and Emulsifying Properties of Sardine (Sardina-Pilchardus) Protein Hydrolysates. *J. Food Sci.* **1990**, *55*, 1571–1573.
20. Schroder, A.; Berton-Carabin, C.; Venema, P.; Cornacchia, L. Interfacial properties of whey protein and whey protein hydrolysates and their influence on O/W emulsion stability. *Food Hydrocoll.* **2017**, *73*, 129–140.
21. Tan, Y.; Chang, S.K.C.; Meng, S. Comparing the kinetics of the hydrolysis of by-product from channel catfish (Ictalurus punctatus) fillet processing by eight proteases. *LWT* **2019**, *111*, 809–820.
22. Aspevik, T.; Oterhals, Å.; Ronning, S.B.; Altintzoglou, T.; Wubshet, S.G.; Gildberg, A.; Afseth, N.K.; Whitaker, R.D.; Lindberg, D. Valorization of Proteins from Co- and By-Products from the Fish and Meat Industry. *Topics Curr. Chem.* **2017**, *375*. [CrossRef]
23. Liceaga-Gesualdo, A.M.; Li-Chan, E.C.Y. Functional properties of fish protein hydrolysate from herring (Clupea harengus). *J. Food Sci.* **1999**, *64*, 1000–1004. [CrossRef]
24. Van der Ven, C.; Gruppen, H.; de Bont, D.B.; Voragen, A.G. Emulsion properties of casein and whey protein hydrolysates and the relation with other hydrolysate characteristics. *J. Agric. Food Chem.* **2001**, *49*, 5005–5012. [CrossRef] [PubMed]
25. Pearce, K.N.; Kinsella, J.E. Emulsifying properties of proteins—Evaluation of a turbidimetric technique. *J. Agric. Food Chem.* **1978**, *26*, 716–723. [CrossRef]
26. Al-Soufi, W.; Piñeiro, L.; Novo, M. A model for monomer and micellar concentration in surfactant solutions: Application to conductivity, NMR, diffusion, and surface tension data. *J. Colloid Interface Sci.* **2012**, *370*, 102–110. [CrossRef]

27. Liu, X.; Yu, Q.; Song, A.; Dong, S.; Hao, J. Progress in nuclear magnetic resonance studies of surfactant systems. *Curr. Opin Colloid Interface Sci.* **2020**, *45*, 14–27. [CrossRef]
28. ISO 5983-2:2009. *Animal Feeding Stuffs—Determination of Nitrogen Content and Calculation of Crude Protein Content*; Part 2: Block Digestion and Steam Distillation Method; International Organization for Standardization (ISO): Geneva, Switzerland, 2009.
29. Sriperm, N.; Pesti, G.M.; Tillman, P.B. Evaluation of the fixed nitrogen-to-protein (N:P) conversion factor (6.25) versus ingredient specific N:P conversion factors in feedstuffs. *J. Sci. Food Agric.* **2011**, *91*, 1182–1186. [CrossRef]
30. Cohen, S.A.; Michaud, D.P. Synthesis of a fluorescent derivatization reagent, 6-aminoquinolyl-N-hydroxysuccinimidyl carbamate and its application for the analysis of hydrolysate amino-acids via high-performance liquid-chromatography. *Anal. Biochem.* **1993** *211*, 279–287. [CrossRef]
31. Aspevik, T.; Egede-Nissen, H.; Oterhals, Å. A Systematic Approach to the Comparison of Cost Efficiency of Endopeptidases for the Hydrolysis of Atlantic Salmon (Salmo salar) By-Products. *Food Technol. Biotech.* **2016**, *54*, 421–431. [CrossRef]
32. Conway, E.J.; Byrne, A. An absorption apparatus for the micro-determination of certain volatile substances: The micro determination of ammonia. *Biochem. J.* **1933**, *27*, 419–429.
33. Wang-Andersen, J.; Haugsgjerd, B.O. *Improved Analysis for Peptide Size Distribution in Marine Hydrolysates (In Norwegian)*; RUBIN Trondheim, Norway, 2011; ISBN 978-82-7251-926-0.
34. Oterhals, A.; Samuelsen, T.A. Plasticization effect of solubles in fishmeal. *Food Res. Int.* **2015**, *69*, 313–321. [CrossRef]
35. Steinsholm, S.; Oterhals, Å.; Underhaug, J.; Måge, I.; Malmendal, A.; Aspevik, T. Sensory Assessment of Fish and Chicken Protein Hydrolysates. Evaluation of NMR Metabolomics Profiling as a New Prediction Tool. *J. Agric. Food Chem.* **2020**, *68*, 3881–3890 [CrossRef] [PubMed]
36. BRENDA. The Comprehensive Enzyme Information System. Available online: http://www.brenda-enzymes.org/ (accessed on 16 September 2019).
37. Kousoulaki, K.; Olsen, H.J.; Albrektsen, S.; Langmyhr, E.; Mjos, S.A.; Campbell, P.; Aksnes, A. High growth rates in Atlantic salmon (Salmo salar L.) fed 7.5% fish meal in the diet. Micro-, ultra- and nano-filtration of stickwater and effects of different fractions and compounds on pellet quality and fish performance. *Aquaculture* **2012**, *338*, 134–146. [CrossRef]
38. Wouters, A.G.B.; Rombouts, I.; Fierens, E.; Brijs, K.; Delcour, J.A. Relevance of the Functional Properties of Enzymatic Plant Protein Hydrolysates in Food Systems. *Compr. Rev. Food Sci. Food Saf.* **2016**, *15*, 786–800. [CrossRef]

foods

MDPI

Article

Prebiotic Potential and Anti-Inflammatory Activity of Soluble Polysaccharides Obtained from Soybean Residue

Bao Le [1], Thi Ngoc Anh Pham [2] and Seung Hwan Yang [2,*]

[1] Faculty of Pharmacy, Ton Duc Thang University, Ho Chi Minh City 700000, Vietnam; lebao@tdtu.edu.vn
[2] Department of Biotechnology, Chonnam National University, Yeosu 59626, Korea; 197621@jnu.ac.kr
* Correspondence: ymichigan@jnu.ac.kr; Tel.: +82-61-659-7306

Received: 6 November 2020; Accepted: 5 December 2020; Published: 6 December 2020

Abstract: In the present study, we assessed the extraction of low molecular weight soluble polysaccharides (MESP) from soybean by-products using microwave-assisted enzymatic technology and proposed the chemical structure of MESP using Fourier transform-infrared spectroscopy, gas chromatography, and ^{1}H and ^{13}C nuclear magnetic resonance spectrum analysis. The results suggested that MESP mainly comprised arabinose, rhamnose, and glucuronic acid with (1→4) glycosidic linkages in the backbone. Compared with inulin, MESP was found to selectively stimulate the growth of *Lactobacillus* probiotics. Moreover, the results of in vitro fermentation indicated that MESP significantly increased the concentrations of both acetate and butyrate ($p < 0.05$). MESP were treated on lipopolysaccharide (LPS)-stimulated RAW264.7 cells to determine the anti-inflammatory effect in vitro. It was observed that MESP inhibited nitric oxide, tumor necrosis factor (TNF)-α, interleukin (IL)-1β, IL-6, and IL-10 production. Furthermore, Western blotting results indicated that MESP significantly attenuated LPS-induced downregulation of phosphorylation levels of Janus kinase 2 (JAK2) and signal transducer and activator of transcription 3 (STAT3) in macrophages. The underlying mechanism might involve inhibition of the expression of pro-inflammatory cytokines, presumably via JAK2/STAT3 pathway. Collectively, the results of our study paved way for the production of MESP, which may be potentially used as nutraceutical ingredients for prebiotics and anti-inflammatory agents, from soybean residue.

Keywords: fermentation; inflammation; microwave-assisted enzymatic extraction; nutraceutical; polysaccharide; prebiotics; probiotics; soybean residue

1. Introduction

Soy (*Glycine max*) is an economical crop that has been cultivated in eastern Asian countries since ancient times, and more recently, is increasingly used in the Western countries [1]. Soybean is used for human consumption, since it is an abundant source of protein and oil. Soymilk consumption has increased in most countries; however, increased soymilk production has been continuously accompanied by soybean by-product accumulation. The residue left from soybean after aqueous extraction from soymilk production is rich in insoluble fiber (55%) and remaining protein (30%) [2]. Several studies have been conducted on soybean by-products to identify different methodologies to minimize the economic and environmental load of food production.

For decades, research has revealed that polysaccharides possess immense health beneficial physiological effects including immunomodulatory [3], antiviral [4], and antioxidant activities [5]. Soybean polysaccharide and its derivatives were reported to possess hypolipemic, hypocholesterolemic, anti-inflammatory, antidiabetic, antiobesity, anticardiovascular disease activities, and these derivatives improved metabolic syndrome. These derivatives may be related to polysaccharide activity [6–8];

however, soybean residue is still considered a waste product due to its low water solubility. Moreover, the correlation between the molecular weight and health benefits of polysaccharides has already been reported [7].

A decrease in the molecular weight of honey-processed *Astragalus* polysaccharide (dextran) was accompanied by its reduced anti-inflammatory activity [9]. A similar trend was observed on investigating a polysaccharide obtained from *Schizophyllum commune* [10]. He et al. [11] observed that low molecular weight polysaccharide might reach more distal regions of the colon intact and may lead to slower fermentation rate, which results in high prebiotic potency. Therefore, it is necessary to find alternative strategies to obtain low molecular weight polysaccharides with high solubility and anti-inflammatory activity from soybean residue.

Several modification methods, such as acid hydrolysis [12], autoclaving [13], and enzymatic hydrolysis [14], have been developed to prepare soluble polysaccharides. Enzymatic hydrolysis used for production of valuable polysaccharides has been intensively studied due to the improved extraction yield of water-soluble polysaccharides and the simple process; however, this method requires strict temperature and pH conditions and is time consuming [15]. In contrast, in our previous study, microwave-assisted extraction (MAE) required shorter time, reduced solvent consumption, and lower energy input [16]. Nevertheless, the effects of the combined use of enzymatic hydrolysis and MAE on the bioactivities of soybean-derived soluble polysaccharides, particularly those with low molecular weight, have not yet been investigated.

Hence, we evaluated the physicochemical characteristics and bioactivities of low molecular weight soluble polysaccharides (MESP) derived from soybean using microwave assisted enzymatic extraction (MAEE). MESP were isolated and purified via a Cellulose DEAE-52 column chromatography. The in vitro chemical characteristics and biological activities, including prebiotic and anti-inflammatory effects, of MESP were investigated.

2. Materials and Methods

2.1. Materials

The solid soy residue obtained from CJ Food Systems (Guro, Seoul, Korea) and was stored in a freezer at −4 °C before use. Inulin from chicory, lipopolysaccharide (LPS) from *Escherichia coli* O111:B4, pectinase from *Aspergillus aculeatus*, cellulase from *A. niger* were provided by Sigma Chemical (St. Louis, MO, USA). MRS carbohydrate-free broth (MB-M0757) and MRS broth (MB-M1025) were purchased from MB cell (Los Angeles, CA, USA). *Lactobacillus rhamnosus* GG ATCC 53103, *L. plantarum* ATCC 8014, *L. brevis* ATCC 367, and murine macrophage cell line RAW 264.7 ATCC TIB71 were purchased from the American Type Culture Collection. Penicillin-Streptomycin (10,000 U/mL), Dulbecco's modified eagle's medium, fetal bovine serum (FBS) was purchased from Gibco BRL (Grand Island, NY, USA). Bicinchoninic acid assay kit (23225) was purchased from Thermo Scientific^{TM} (Waltham, MA, USA). Griess reagent kit (G7921), tumor necrosis factor (TNF-α) (A43658), IL-6 (P26892), IL-10 (P18893), and IL-1β (P10749) were purchase from Invitrogen (Carlsbad, CA, USA). Mouse anti-β-actin (sc-47778), Janus kinase 2 (JAK2) (sc-390539), phosphorylated (p)-JAK2 (sc-101718), signal transducer and activator of transcription 3 (STAT3) (sc-8019), and p-STAT3 (sc-8059) antibodies were purchased from Santa Cruz Biotechonlogy (Santa Cruz, CA, USA). Anti-rabbit horseradish peroxidase-conjugated secondary antibodies (ab6721) was acquired from Abcam (Cambridge, MA, USA). All chemicals and reagents applied were of analytical grade.

2.2. Microwave Assisted Enzymatic Extraction

The fresh soybean residue was dried at 60 °C for 3 d and followed by grounding (80 mesh) in a FM-909 T mechanical grinder (Hanil Electric Co., Seoul, Korea) to obtain a homogeneous soybean residue powder. The powder was pretreated with 80% ethanol at 60 °C for 5 h to remove oligosaccharide, colored materials, and some small molecule materials. The insoluble parts were separated via centrifugation and dried until its weight was constant. The microwave extraction was performed as previously described by Le, Golokhvast, Yang, and Sun [16] with minor modifications. The pretreated powder of 10 g was accurately weighed and was immersed in 200 mL deionized water (D_2O) (1:10 *w/v*) for 1 h in a 250 mL flat-bottomed flask. The mixture was hydrolyzed for 30 min using a cocktail enzyme (pectinase: cellulase = 1:1, mixed enzymes: powder = 2.0%, *w/v*) under the microwave irradiation power of 700 W. Thereafter, the suspension was centrifuged and concentrated to reach 20 mL by rotary evaporation under reduced pressure. Next, the mixture was precipitated with 80% (*v/v*) ethanol) at 4 °C overnight. The crude polysaccharide was obtained via centrifugation, deproteinized by the Sevag reagent (chloroform: n-butyl alcohol = 4:1, *v/v*) to remove the associated proteins, and finally lyophilized through freeze drying process. The crude polysaccharide was redissolved in distilled water and fractionated through an activated Cellulose DEAE-52 (5.0 × 30 cm) column pre-equilibrated with distilled water and was then stepwise eluted with distilled water and 0.5 M NaCl at a flow rate of 1.0 mL/min (10 mL/tube). Each fraction was evaluated at 490 nm by phenol sulfuric acid to detect the compounds. The main fraction was extensively dialyzed against ultrapure water at 4 °C for 48 h (molecular weight cut-off: 3.5 kDa), and lyophilized to obtain white pure polysaccharide, namely MESP, for further analyses.

2.3. Characterization of MESP

Total polysaccharide content was measure by the phenol-sulfuric acid colorimetric method at 621 nm using D-glucose as a standard. The molecular weight of MESP was measured according to a method described by Zheng et al. [17]. Monosaccharide composition of MESP was determined via a gas chromatography-mass spectrometer (6890/5973N-GC/MSD, Agilent Technologies, Santa Clara, CA, USA) after hydrolysis with 3 M trifluoroacetic acid at 100 °C for 3 h, and then the MESP were passed through a 0.45 µm syringe filter before loading on a capillary column HP-5 ms (30 m × 0.25 mm × 0.25 µm) at 25 °C. VERTEX 70 v Fourier transform infrared spectrometer (Bruker, Germany) was used for FT-IR measurements. A potassium bromide disc that contains 1% of MESP was recorded in the frequency range of 4000–400 cm^{-1}. MESP (30 mg) was maintained over P_2O_5 under vacuum for several days and then dissolved in D_2O. The solid-state ^1H and ^{13}C nuclear magnetic resonance (NMR) spectra were recorded at 25 °C on a Varian Inova 500 and 600 MHz spectrometers (Varian, Walnut Creek, CA, USA), which were operated at 400 MHz and 101 MHz for ^1H and ^{13}C NMR, respectively.

2.4. Stimulation of Probiotic Growth

The prebiotic activity of MESP was evaluated by a method described in a previous study with minor modifications [18]. MRS carbohydrate-free broth (MB cell, Los Angeles, CA, USA) was used as the basal medium, and MESP and inulin (positive prebiotic control) were used as the individual carbon sources. Three *Lactobacillus* strains, including *L. rhamnosus* GG, *L. plantarum*, and *L. brevis*, were precultured in MRS broth, centrifuged and diluted in basal MRS medium to attain 1 × 10^7 colony forming units (CFU)/mL. The MRS carbohydrate-free broth containing 1% carbon source (MESP or inulin) was dispensed as 200 µL aliquots into in 96-well U-shaped-bottom microplate; thereafter, 20 µL of logarithmic culture of each probiotic strains was added for fermentation. The growth of the probiotics was evaluated after 24 h of incubation using an ELISA reader (BioTek Instruments, Winooski, VT, USA) at a wavelength of 600 nm.

2.5. Short Chain Fatty Acid (SCFA) Analysis

In vitro fermentation was performed according to a previously described method [19] with certain modifications. Fresh feces samples were obtained from three healthy individuals (2 female, 1 male, aged 22–29 years), who had not been treated with antibiotics in the last 3 months. Feces were homogenized with 0.1 M anaerobic phosphate-buffered saline (pH 7.0) using a Bead Ruptor Homogenizer (Omni International, NW Kennesaw, GA, USA) to make 10% (*w/v*) slurries. The basic medium comprised the followings components (per liter): 0.45 g KH_2PO_4; 0.45 g K_2HPO_4; 0.05 g NaCl;

0.064 g KCl; 1.0 g $MgSO_4 \cdot 7H_2O$; 0.5 g cysteine-HCl; 0.5 g bile salts, 0.05 g hemin; 0.001 g resazurin; 2.0 mL Tween 80; 0.01 mL vitamin K, with or without supplement of 8.0 g each of MESP or inulin. The fermentation was carried out in an anaerobic chamber at 37 °C for 24 h. After fermentation, the culture (1 mL) was dispensed into a sterile Eppendorf tube (1.5 mL) and centrifuged at 13,000× *g* for 10 min to collect the supernatant. Next, the supernatants were filtered using organic phase microfiltration membrane (0.22 mm) and analyzed by Agilent 1100 Series (Agilent Technologies, Inc., Santa Clara, CA, USA) with a DAD detector at 210 nm equipped with a calcium-loaded Aminex-HPX-87C column (Bio-Rad Corp., Richmond, CA, USA).

2.6. Cytotoxicity Assay

We examined the cytotoxicity of MESP on RAW 264.7 murine macrophage cells by growing them in Dulbecco's modified eagle's medium supplemented with combination of penicillin and streptomycin (final concentration of 100 units/mL and 100 µg/mL, respectively), along with 10% fetal bovine serum at 37 °C, with 5% CO_2 supply. As the cells reached 80% confluence, they were treated with different concentrations of MESP (5–1280 µg/mL) for 24 h. The cell vitality was measured using methylthiazolyldiphenyl-tetrazolium bromide (MTT) assay using microplate reader (F50, Tecan, Mannedorf, Switzerland) via spectrophotometry at 570 nm.

2.7. Measurement of NO and Cytokines

To evaluate anti-inflammatory potential of MESP, RAW264.7 cells (5×10^5 cells/mL) were seeded in a 24-well plate and adding MESP at various concentrations in the presence of LPS (1 µg/mL) in 5% CO_2 at 37 °C for 24 h. The NO, TNF-α, IL-6, IL-10, and IL-1β concentrations in the supernatant were measured by Griess and ELISA assays according to the manufacturer's instructions.

2.8. Western Blot Analysis

RAW264.7 cells were rinsed with cold 1× phosphate buffer saline and detached with EDTA-free 0.25% trypsin, followed by homogenization using RIPA lysis buffer (Sigma-Aldrich, St. Louis, MO, USA). After quantitative determination of the protein content by BCA assay kit, equal quantities of total proteins were separated by denaturing them via 10% SDS-PAGE and then transferred to PVDF membranes (GE Healthcare, Buckinghamshire, UK). The membranes were then incubated in Tris-buffered saline containing 0.1% (*v/v*) Tween-20 (TBST) and 5% skim milk at 37 °C for 3 h, and the specific primary antibodies were incubated at 4 °C overnight. After several washes with TBST (0.075%), they were incubated with anti-rabbit horseradish peroxidase-conjugated secondary antibodies for 1 h at room temperature. The targeted protein blots were developed using an enhanced chemiluminescence (ECL) kit (ECL-plus, Thermo Scientific, USA) and detected by using a Chemiluminometer (CLINX Scientific Instrument Co., Ltd., Shanghai, China).

2.9. Statistical Analysis

All results were performed in triplicate and expressed as mean ± standard deviation. Analysis of variance and Student's t-test were analyzed using the SPSS software (IBM, SPSS 22.0, Chicago, IL, USA) at $p < 0.05$ and $p < 0.01$.

3. Results

3.1. Polysaccharide Extraction, Isolation, and Purification

Four fractions were obtained by NaCl (0 and 0.5 M) elution (Figure 1a). Each polysaccharide peak was collected, dialyzed, and lyophilized. Due to the lower yields observed for other eluted fractions (1.72–11.41%), the main MESP fraction (80.86%) was selected for structural and functional characterizations. The single and symmetrically sharp peak at 19.02 min in the MESP profile indicated that MESP was a homogeneous polysaccharide with molecular weight of 3.32 kDa, based on the calibration curve of dextran (Log $M_W = -0.3508 \times$ retention time + 10.193; Figure 1b). The total sugar content of MESP was 97.69% with remaining trace amounts of protein and nucleic acid.

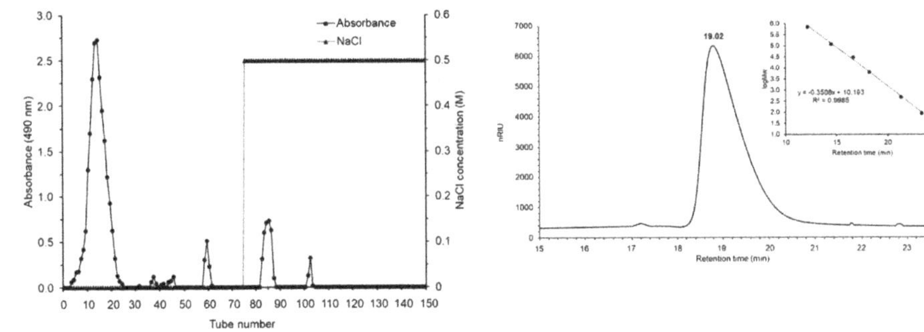

(a) (b)

Figure 1. Elution curve obtained using Sephadex G-75 gel filtration (**a**) and HPLC chromatograms (**b**).

Gas chromatography (GC) analysis of the monosaccharides (Table 1) revealed that MESP comprised arabinose, galacturonic acid, rhamnose, xylose, galactose, glucose, and fucose at a molar ratio of 27.4:18.1:11.41:5.71:2.53:1.53:1.32, indicating that arabinose, galacturonic acid, and rhamnose were the main monosaccharide components of low molecular weight soluble polysaccharides (MESP).

Table 1. Monosaccharide composition of okara polysaccharide.

Sample	Molecular Weight (kDa)	Monosaccharide Compositions (mol%)							
		Ara	Rha	Fuc	Xyl	Man	Gal	Glc	GalA
Soybean residue		16.41	3.46	0.91	12.89	ND	27.56	10.23	20.3
MESP	3.4	27.4	11.41	1.32	5.71	ND	2.53	1.53	18.1

3.2. Structural Characterization

The integrated structure of MESP was further assessed via FT-IR and ^{1}H and ^{13}C NMR. The typical FT-IR spectrum of MESP in the 400–4000 cm^{-1} range is presented in Figure 2a. The graft MESP revealed strong stretching peak at 3397.8 cm^{-1} corresponding to the hydroxyl groups (O-H) [20]. The weak absorption peak at 2919.2 cm^{-1} was characteristic of C-H asymmetric stretching vibration [21]. The bands at 1736.2 and 1629.6 cm^{-1} might be due to the nonsymmetrical vibration of C=O in the ester group [22]. Moreover, the absorption signals at around 1473.3 and 1074.1 cm^{-1} indicated the presence of residual water (OH^-). Furthermore, the signal observed at 1256.3 cm^{-1} was attributed to the stretching vibration of C-O. Collectively, FT-IR analysis suggested that MESP possessed typical sugar groups.

The ^{1}H and ^{13}C NMR spectra, illustrated in Figure 2b,c, confirmed the attribution and structural characterization of MESP. Based on the data in existing literature, the signal of five anomeric protons at δ3.18–4.18 was identified as the zone of accumulation of protons from α-D-GalA$^-$ residues, whereas that at δ1.21–4.03 was assigned to α-L-Rha*p*$^-$ [8]. The ^{13}C NMR spectrum of MESP showed no signal at low field from 160 to 180 ppm, indicating that MESP did not contain uronic acid. Chemical shifts of MESP are summarized in Table 2. According to the corresponding chemical shifts in existing literature, MESP structure was proposed, as shown in Figure 2d.

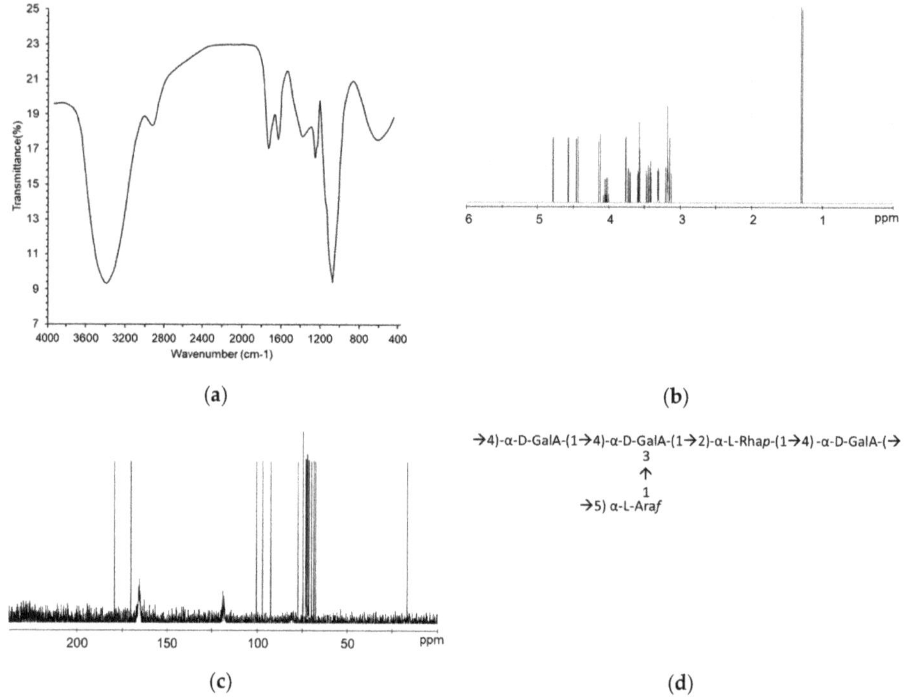

Figure 2. Fourier transform infrared spectra (**a**); ^{1}H-NMR spectra (**b**); ^{13}C-NMR spectra (**c**); and proposed structure (**d**) of low molecular weight soluble polysaccharides (MESP).

Table 2. Summary of ^{1}H and ^{13}C chemical shifts for MESP.

Residues			**Chemical Shift Assignment of MESP**					
			1	**2**	**3**	**4**	**5**	**6**
A	→4)-α-D-GalA-(1→	H	4.58	3.18	3.58	3.42	3.76	
		C	94.3	74.0	74.9	73.2	76.4	172.2
B	→2)-α-L-Rha*p*-(1→	H	4.53	3.57	3.46	3.15	4.03	1.21
		C	99.0	79.2	69.4	71.7	70.4	18.3
C	α-D-GalA-(1→	H	4.45	3.18	3.33	3.71	4.14	
		C	98.4	74.1	74.2	73.5	74.6	181.1
D	→5) α-L-Ara*f* (1→	H	4.88	3.56	3.52	3.37	3.95	
		C	98.59	72.2	69.6	71.23	61.01	15.9

3.3. MESP Stimulated Probiotic Growth

The number of all tested *Lactobacillus* strains treated with MESP or inulin was remarkably increased (Figure 3a), indicating that all *Lactobacillus* strains could utilize MESP and inulin. *Lactobacillus* exhibited an outstanding growth rate when MESP, in contrast to inulin, was used as the carbon source.

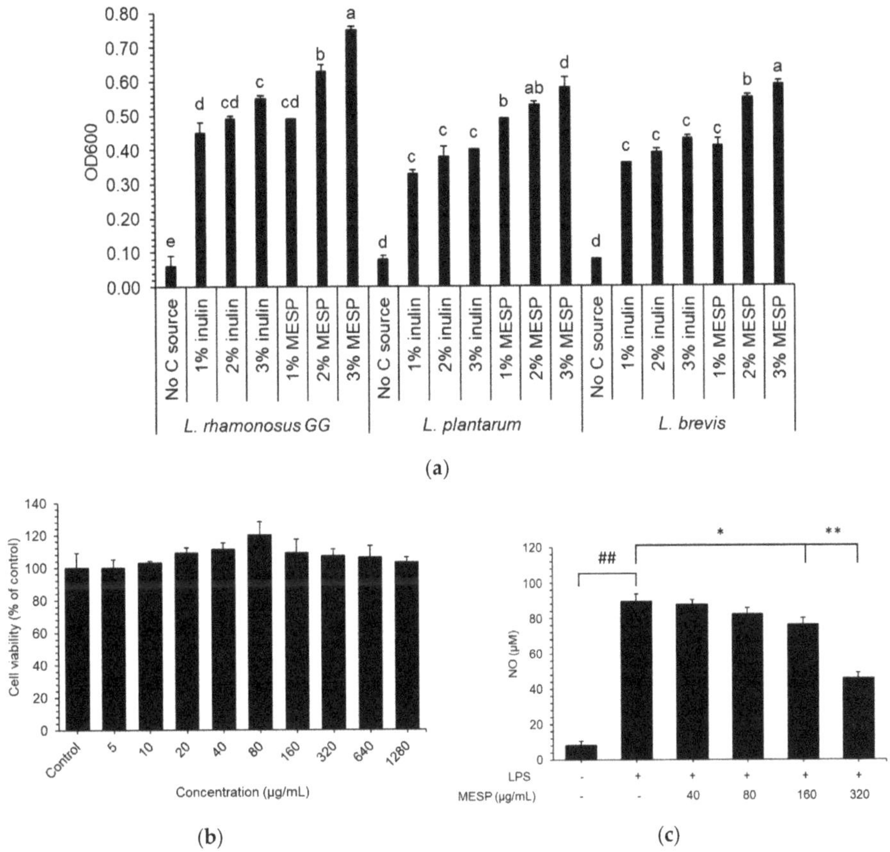

Figure 3. Growth of probiotics in glucose free-MRS base media with supplement of inulin or MESP (**a**). OD600, optical density of a sample measured at a wavelength of 600 nm. Values are expressed as mean ± SD ($n = 3$). Different letters indicate significant ($p < 0.05$) differences in the same strain. Cytotoxic effects (**b**) and NO production (**c**) of MESP in RAW 264.7 cells. ## $p < 0.01$, compared to the control group; * $p < 0.05$, ** $p < 0.01$, compared to the lipopolysaccharide (LPS) treated group.

3.4. Effect of MESP on SCFA Production

The pH of the cecum is lower than that of the ileum, which in turn affects the growth inhibition of pH-sensitive pathogenic bacteria, thereby altering gut microbiota composition and promoting host health. After in vitro fermentation, the concentrations of total short chain fatty acids (SCFAs), including acetic, butyric, and propionic acids, were significantly higher ($p < 0.05$) in the MESP group than those in the control group (Table 3), indicating that MESP enhanced SCFA production in the gut model. A comparison of the results revealed that the concentrations of total SCFAs were significantly higher ($p < 0.05$) after MESP treatment than those after inulin treatment. Acetic acid concentration was significantly increased ($p < 0.05$).

Table 3. Concentrations of short chain fatty acids (SCFAs) at different points during in vitro fermentation.

SCFAs (mM)	Sample	Fermentation Time (h) 0	6	12	24	48
Acetic acid	Control	3.89 ± 0.48 H [1]	4.29 ± 0.69 cH	16.48 ± 0.71 bF	15.39 ± 0.67 cF	15.48 ± 0.99 cF
	1% inulin		9.48 ± 0.49 bG	21.86 ± 0.89 bE	29.82 ± 0.08 bC	24.36 ± 0.53 bD
	1% MESP		24.18 ± 0.36 aD	51.84 ± 0.05 aB	69.48 ± 0.74 aA	69.14 ± 0.62 aA
Butyric acid	Control	2.18 ± 0.31 E	2.95 ± 0.12 bE	4.85 ± 0.49 bD	8.62 ± 0.18 bB	6.21 ± 0.83 cC
	1% inulin		4.31 ± 0.81 aD	6.14 ± 0.48 aC	9.89 ± 0.84 bB	9.18 ± 0.23 bB
	1% MESP		4.18 ± 0.69 aD	6.83 ± 0.41 aC	13.21 ± 0.55 aA	13.23 ± 0.81 aA
Propionic acid	Control	1.22 ± 0.23 D	1.33 ± 0.04 bD	2.89 ± 0.63 bD	4.26 ± 0.47 bC	4.11 ± 0.84 bC
	1% inulin		4.15 ± 0.28 aC	6.81 ± 0.16 aB	11.51 ± 0.04 aA	11.81 ± 0.41 aA
	1% MESP		4.92 ± 0.85 aC	7.21 ± 0.74 aB	12.21 ± 0.56 aA	12.18 ± 0.07 aA
Total SCFAs	Control	9.12 ± 0.11 G	9.34 ± 0.75 cG	22.14 ± 0.44 cE	34.95 ± 0.12 bD	32.43 ± 0.64 cD
	1% inulin		11.15 ±0.07 bF	33.81 ± 0.49 bD	45.12 ± 0.17 bC	44.26 ± 0.16 bC
	1% MESP		29.81 ± 0.09 aE	63.31 ± 0.18 aB	83.49 ± 0.78 aA	82.46 ± 0.36 aA

[1] Values are expressed as means ± SD (n = 3). Values followed by the same capital letters are not significantly different at $p < 0.05$ among different samples at the same time point, while the same minuscules letter are not significantly different among different times ($p < 0.05$) in the same SCFAs.

3.5. *Effects of MESP on RAW264.7 Cell Proliferation*

We analyzed the safe concentration of MESP on RAW 264.7 macrophage, with various MESP concentrations wherein cell viability was determined by MTT assay. We observed that MESP at concentrations 5–1280 μg/mL did not affect cell growth in a dose-dependent manner. (Figure 3b).

3.6. *Effects of MESP on NO Production in LPS-Stimulated RAW264.7 Cells*

The NO levels were markedly induced after treatment with LPS alone, whereas they were remarkably inhibited by MESP pretreatment in a dose-dependent manner (Figure 3c). As 160 μg/mL MESP did not present any cytotoxic effect, we propose the possibility that inhibition of NO production was not linked to the cytotoxic effects on RAW264.7 cells.

3.7. *Effect of MESP on IL-6, TNF-α, and IL-1β Release in LPS-Stimulated RAW264.7 Cells*

As illustrated in Figure 4, inflammatory cytokine production was tremendously increased in the LPS-treated group compared with that in the control group ($p < 0.01$), whereas TNF-α, IL-6, IL-1β, and IL-10 production was significantly inhibited by the six tested peptides in LPS-treated RAW264.7 cells, compared with that in the control cells. Treatment with MESP at concentration 20 μg/mL showed the strongest inhibitory activity towards TNF-α, IL-6, IL-1β, and IL-10 at 57.2%, 30.1%, 45.7%, and 55.3%, respectively.

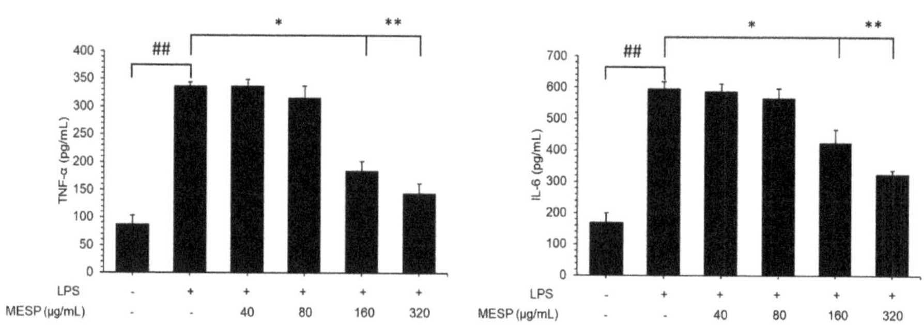

(a) **(b)**

Figure 4. *Cont.*

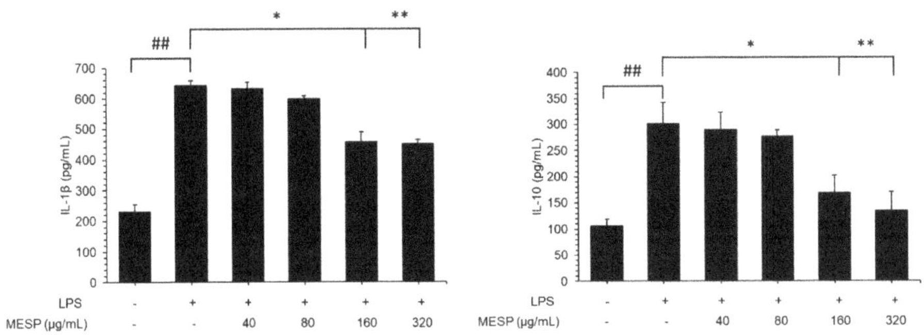

(c) (d)

Figure 4. Inhibitory effects of MESP on TNF-α (**a**), IL-6 (**b**), IL-1β (**c**), IL-10 (**d**) production on RAW 264.7 cells by ELISA. Values are expressed as mean ± SD ($n = 3$). ## $p < 0.01$, compared to the control group; * $p < 0.05$, ** $p < 0.01$, compared to the LPS treated group.

3.8. MESP Blocked the LPS-Triggered Inflammatory Response via the JAK2/STAT3 Pathway

We revealed that LPS induced JAK2 and STAT3 phosphorylation (Figure 5). Next, we explored the role of MESP in JAK2 and STAT3 regulation. As illustrated in Figure 5, MESP could significantly suppress phosphorylated (LPS-induced) JAK2/STAT3, indicating that MESP exerted anti-inflammatory effects through the JAK2/STAT3 signaling pathway.

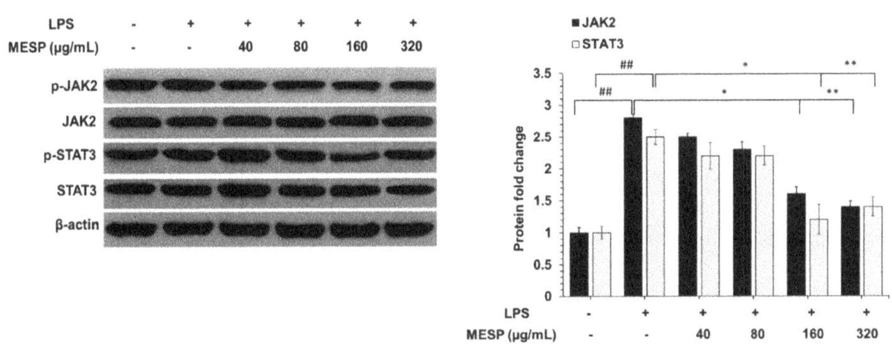

(a) (b)

Figure 5. Inhibitory effects of MESP on JAK2/STAT3 pathway activation. Immunoblot analysis of JAK2 and STAT3 protein levels (**a**) and quantification of protein levels (**b**) normalized to β-actin. ## $p < 0.01$, compared to the control group; * $p < 0.05$, ** $p < 0.01$, compared to the LPS treated group.

4. Discussion

In the present study, crude polysaccharide was extracted from soybean residue by MAEE extraction and then fractionated using the Cellulose DEAE-Sephadex A-25 column. The extraction yield was 73.8%, which is comparable to that (previously reported value ranging from 7.09–85%) obtained using alkaline hydrogen peroxide, enzymatic hydrolysis, autoclaving, fermentation, and ultrasonic-assisted extraction [23–25]. In MAEE, the extraction time was shorter than that in microwave-assisted extraction (0.5–2 h), whereas the extraction temperature was lower than that for autoclaving (121 °C); moreover, enzyme loading of MAEE extraction was lower than that of enzymatic hydrolysis using cellulase and pectinases, thereby suggesting that MAEE extraction was more efficient and eco-friendly. In a previous study, numerous heterogeneous polysaccharides isolated from soybean residue (okara) presented monosaccharide compositions comprising mainly galacturonic acid, galactose, arabinose, xylose, fucose, and rhamnose [2,26]. Furthermore, enzyme treatment increased arabinose content in polysaccharides obtained from tea leaf and pulp [27,28], and in accordance with our results. This demonstrated that MESP was a pectin-like polysaccharide with rhamnose and galacturonic acid forming its backbone.

To evaluate the growth stimulating effects of MESP on probiotics, MESP was used as the single carbon source for *Lactobacillus* spp. cultivation. Meanwhile, the medium without a carbon source was used as a negative control, whereas inulin (commercial prebiotic) was used as a positive control to compare its effects with those of MESP. Our results were in accordance to those of previous in vitro and in vivo studies, which revealed that soybean oligosaccharides and soluble polysaccharides could stimulate *Lactobacillus* and *Bifidobacterium* growth [29–31]. Soybean soluble polysaccharides, which are rich in GalA and Ara, were reported to stimulate *Lactobacillus fermentum* and *L. plantarum* growth by 49.5 and 5.25 times, respectively [29]. A similar result was observed, demonstrating that consumption of high-dose soybean oligosaccharides could promote *Bifidobacterium* and *Lactobacillus* proliferation in a mouse model [32]. This indicated that MESP was a good substrate for facilitating probiotic growth.

SCFAs are the main end-products obtained fermentation of dietary fiber by gut microbes in the colons [19]. This increase was consistent with the change in the number of *Lactobacillus* spp. (known as lactate producers). Enriched acetate production has been reported in numerous inulin intervention studies [14,29]. Moreover, a similar result has been documented when soybean oligosaccharides were used. Zhou, et al. [33] found that the fermentation of mini-pig-supplemented soybean oligosaccharides increased acetic, propionic, and butyrate acid production, and this could be corrected by increasing the number of *Lactobacillus* spp. Our results indicated that MESP exerted better effects on SCFA production and richness than inulin, and this might be due to the more diverse monosaccharide compositions of MESP.

Macrophages are crucial immunocytes and play a pivotal role in protecting the host from certain pathogens and maintaining homeostasis [34]. Our result revealed that MESP exerted no toxicity toward RAW264.7 cells, and this was in accordance with the results of previous studies [32,35]. Moreover, soybean oligosaccharide was considered "generally recognized as safe" material in the United States. To estimate the suppressive effects on NO, pro-inflammatory mediator in LPS-induced inflammation, RAW264.7 cells were treated with MESP at serial concentrations for 4 h before treatment with LPS (1 μg/mL). These data suggested that MESP exerts anti-inflammatory activity towards RAW264.7 cells. To determine the underlying anti-inflammatory mechanisms of MESP, we evaluated the production of inflammatory cytokines, which are crucially involved in the inflammatory response. The immunomodulatory capacities were affected by the different monosaccharide compositions of the MESP, and this was in accordance with the results of a previous study [35]. Furthermore, the (1→4) glycosidic bond, which is the main type of linkage formed in MESP, might partly be responsible for the immunoregulatory activity of MESP [36,37]. A previous study showed that soybean curd residue at higher concentrations (800–1600 μg/mL) significantly downregulated LPS-induced TNF-α and IL-1β protein expression [38]. Additionally, admixtures of soybean soluble polysaccharides and genistein reduced IL-1, IL-6, and TNF-α levels in high-dose L-carnitine-fed mice [39]. Recent studies have confirmed the importance of JAK2/STAT3 in the macrophage inflammatory response [40,41]. STAT3 is a protein that mediates the expression of numerous cytokines such as IL-6 to promote inflammation. p-STAT3 can subsequently translocate into the nucleus and bind to DNA response elements, and it is generally activated by regulating the activity of the cytoplasmic tyrosine kinase JAK2 [42]. Further evidence that the JAK2/STAT3 signaling pathway could be activated by a series of cytokines, including TNF-α, IL-1β, and IL-6, has been presented [43].

5. Conclusions

The present study revealed that the use of MAE extraction, followed by anion exchange chromatography, allowed the production of low molecular weight (3.32 kDa) MESP from soybean residue. MESP mainly comprised galacturonic acid, rhamnose, and arabinose with copious (1→4) glycosidic linkages in its backbone. MESP exhibited remarkable in vitro prebiotic properties, including stimulation of *Lactobacillus* growth and SCFA production. Furthermore, MESP possessed potential anti-inflammatory activity that might be mediated by inhibiting NO, IL-6, IL-1β, IL-10, and TNF-α production by suppressing the JAK2/STAT3 pathway in LPS-stimulated RAW264.7 cells.

Author Contributions: Conceptualization, methodology, B.L.; formal analysis, validation, writing—original draft preparation, T.N.A.P.; supervision, project administration, S.H.Y. All authors have read and agreed to the published version of the manuscript.

Funding: This research received no external funding.

Acknowledgments: This research was financially supported by the Ministry of Small and Medium-sized Enterprises (SMEs) and Startups (MSS), Korea, under the "Regional Specialized Industry Development Plus Program (R&D, S3005783)" supervised by the Korea Institute for Advancement of Technology (KIAT).

Conflicts of Interest: The authors declare no conflict of interest.

References

1. Yamashita, Y.; Sakakibara, H.; Toda, T.; Ashida, H. Insights into the potential benefits of black soybean (*Glycine max* L.) polyphenols in lifestyle diseases. *Food Funct.* **2020**, *11*, 7321–7339. [CrossRef] [PubMed]
2. Colletti, A.; Attrovio, A.; Boffa, L.; Mantegna, S.; Cravotto, G. Valorisation of by-products from soybean (*Glycine max* (L.) Merr.) processing. *Molecules* **2020**, *25*, 2129. [CrossRef] [PubMed]
3. Guarino, M.P.L.; Altomare, A.; Emerenziani, S.; Di Rosa, C.; Ribolsi, M.; Balestrieri, P.; Iovino, P.; Rocchi, G.; Cicala, M. Mechanisms of action of prebiotics and their effects on gastro-intestinal disorders in adults. *Nutrients* **2020**, *12*, 1037. [CrossRef] [PubMed]
4. Best, T.; Howe, P.; Bryan, J.; Buckley, J.; Scholey, A. Acute effects of a dietary non-starch polysaccharide supplement on cognitive performance in healthy middle-aged adults. *Nutr. Neurosci.* **2015**, *18*, 76–86. [CrossRef] [PubMed]
5. Qiu, S.; Chen, J.; Chen, X.; Fan, Q.; Zhang, C.; Wang, D.; Li, X.; Chen, X.; Chen, X.; Liu, C. Optimization of selenylation conditions for lycium barbarum polysaccharide based on antioxidant activity. *Carbohydr. Polym.* **2014**, *103*, 148–153. [CrossRef]
6. Bai, Z.; Meng, J.; Huang, X.; Wu, G.; Zuo, S.; Nie, S. Comparative study on antidiabetic function of six legume crude polysaccharides. *Int. J. Biol. Macromol.* **2020**, *154*, 25–30. [CrossRef]
7. Salarbashi, D.; Bazeli, J.; Tafaghodi, M. Environment-friendly green composites based on soluble soybean polysaccharide: A review. *Int. J. Biol. Macromol.* **2019**, *122*, 216–223. [CrossRef]
8. Liu, J.; Wen, X.-Y.; Zhang, X.-Q.; Pu, H.-M.; Kan, J.; Jin, C.-H. Extraction, characterization and in vitro antioxidant activity of polysaccharides from black soybean. *Int. J. Biol. Macromol.* **2015**, *72*, 1182–1190. [CrossRef]
9. Liao, J.; Li, C.; Huang, J.; Liu, W.; Chen, H.; Liao, S.; Chen, H.; Rui, W. Structure characterization of honey-processed *Astragalus* polysaccharides and its anti-inflammatory activity *in vitro*. *Molecules* **2018**, *23*, 168. [CrossRef]
10. Du, B.; Zeng, H.; Yang, Y.; Bian, Z.; Xu, B. Anti-inflammatory activity of polysaccharide from *Schizophyllum commune* as affected by ultrasonication. *Int. J. Biol. Macromol.* **2016**, *91*, 100–105. [CrossRef]
11. He, S.; Wang, X.; Zhang, Y.; Wang, J.; Sun, H.; Wang, J.; Cao, X.; Ye, Y. Isolation and prebiotic activity of water-soluble polysaccharides fractions from the bamboo shoots (*Phyllostachys praecox*). *Carbohydr. Polym.* **2016**, *151*, 295–304. [CrossRef] [PubMed]
12. Yoshii, H.; Furuta, T.; Maeda, H.; Mori, H. Hydrolysis kinetics of okara and characterization of its water-soluble polysaccharides. *Biosci. Biotechnol. Biochem.* **1996**, *60*, 1406–1409. [CrossRef] [PubMed]
13. Kasai, N.; Murata, A.; Inui, H.; Sakamoto, T.; Kahn, R.I. Enzymatic high digestion of soybean milk residue (okara). *J. Agric. Food Chem.* **2004**, *52*, 5709–5716. [CrossRef] [PubMed]
14. Villanueva-Suárez, M.J.; Pérez-Cózar, M.L.; Redondo-Cuenca, A. Sequential extraction of polysaccharides from enzymatically hydrolyzed okara byproduct: Physicochemical properties and *in vitro* fermentability. *Food Chem.* **2013**, *141*, 1114–1119. [CrossRef]
15. Jia, X.; Chen, M.; Wan, J.-B.; Su, H.; He, C. Review on the extraction, characterization and application of soybean polysaccharide. *RSC advances* **2015**, *5*, 73525–73534. [CrossRef]

16. Le, B.; Golokhvast, K.S.; Yang, S.H.; Sun, S. Optimization of microwave-assisted extraction of polysaccharides from *Ulva pertusa* and evaluation of their antioxidant activity. *Antioxidants* **2019**, *8*, 129. [CrossRef]
17. Zheng, W.; Chen, X.; Xue, Y.; Hu, J.; Gao, M.-T.; Tsang, Y.F. The influence of soluble polysaccharides derived from rice straw upon cellulase production by *Trichoderma reesei*. *Process Biochem.* **2017**, *61*, 130–136. [CrossRef]
18. Nowak, R.; Nowacka-Jechalke, N.; Juda, M.; Malm, A. The preliminary study of prebiotic potential of Polish wild mushroom polysaccharides: The stimulation effect on *Lactobacillus* strains growth. *Eur. J. Nutr.* **2018**, *57*, 1511–1521. [CrossRef]
19. Xie, Z.; Wang, S.; Wang, Z.; Fu, X.; Huang, Q.; Yuan, Y.; Wang, K.; Zhang, B. *In vitro* fecal fermentation of propionylated high-amylose maize starch and its impact on gut microbiota. *Carbohydr. Polym.* **2019**, *223*, 115069. [CrossRef]
20. Shen, C.-Y.; Jiang, J.-G.; Li, M.-Q.; Zheng, C.-Y.; Zhu, W. Structural characterization and immunomodulatory activity of novel polysaccharides from *Citrus aurantium* Linn. variant amara Engl. *J Funct Foods* **2017**, *35*, 352–362. [CrossRef]
21. Zhang, C.; Gao, F.; Gan, S.; He, Y.; Chen, Z.; Liu, X.; Fu, C.; Qu, Y.; Zhang, J. Chemical characterization and gastroprotective effect of an isolated polysaccharide fraction from *Bletilla striata* against ethanol-induced acute gastric ulcer. *Food Chem. Toxicol.* **2019**, *131*, 110539. [CrossRef] [PubMed]
22. Gao, H.; Zhang, W.; Wang, B.; Hui, A.; Du, B.; Wang, T.; Meng, L.; Bian, H.; Wu, Z. Purification, characterization and anti-fatigue activity of polysaccharide fractions from okra (*Abelmoschus esculentus* (L.) Moench). *Food Funct.* **2018**, *9*, 1088–1101. [CrossRef] [PubMed]
23. Al Loman, A.; Ju, L.-K. Enzyme-based processing of soybean carbohydrate: Recent developments and future prospects. *Enzyme Microb. Technol.* **2017**, *106*, 35–47. [CrossRef] [PubMed]
24. Chen, W.; Duizer, L.; Corredig, M.; Goff, H.D. Addition of soluble soybean polysaccharides to dairy products as a source of dietary fiber. *J. Food Sci.* **2010**, *75*, C478–C484. [CrossRef] [PubMed]
25. Yamaguchi, F.; Kojima, H.; Muramoto, M.; Ota, Y.; Hatanaka, C. Effects of hexametaphosphate on soybean pectic polysaccharide extraction. *Biosci. Biotechnol. Biochem.* **1996**, *60*, 2028–2031. [CrossRef]
26. Li, B.; Lu, F.; Nan, H.; Liu, Y. Isolation and structural characterisation of okara polysaccharides. *Molecules* **2012**, *17*, 753–761. [CrossRef]
27. Huang, F.; Liu, H.; Zhang, R.; Dong, L.; Liu, L.; Ma, Y.; Jia, X.; Wang, G.; Zhang, M. Physicochemical properties and prebiotic activities of polysaccharides from longan pulp based on different extraction techniques. *Carbohydr. Polym.* **2019**, *206*, 344–351. [CrossRef]
28. Wang, Y.; Yang, Z.; Wei, X. Sugar compositions, α-glucosidase inhibitory and amylase inhibitory activities of polysaccharides from leaves and flowers of *Camellia sinensis* obtained by different extraction methods. *Int. J. Biol. Macromol.* **2010**, *47*, 534–539. [CrossRef]
29. Lin, D.; Long, X.; Xiao, L.; Wu, Z.; Chen, H.; Zhang, Q.; Wu, D.; Qin, W.; Xing, B. Study on the functional properties and structural characteristics of soybean soluble polysaccharides by mixed bacteria fermentation and microwave treatment. *Int. J. Biol. Macromol.* **2020**, *2020*, 561–568. [CrossRef]
30. Hou, G.; Jin, M.; Ye, Z.; Zhang, X.; Huang, Q.; Ye, M. Ameliorate effects of soybean soluble polysaccharide on adenine-induced chronic renal failure in mice. *Int. J. Biol. Macromol.* **2020**, *149*, 158–164. [CrossRef]
31. Ursekar, B.; Soni, P.; Date, A.A.; Nagarsenker, M. Characterization of soy polysaccharide and its *in vitro* and *in vivo* evaluation for application in colon drug delivery. *AAPS PharmSciTech* **2012**, *13*, 934–943. [CrossRef] [PubMed]
32. Ma, Y.; Wu, X.; Giovanni, V.; Meng, X. Effects of soybean oligosaccharides on intestinal microbial communities and immune modulation in mice. *Saudi J. Biol. Sci.* **2017**, *24*, 114–121. [CrossRef] [PubMed]
33. Zhou, X.; Kong, X.; Yang, X.; Yin, Y. Soybean oligosaccharides alter colon short-chain fatty acid production and microbial population *in vitro*. *J. Anim. Sci.* **2012**, *90*, 37–39. [CrossRef] [PubMed]
34. Zhang, L.; Wang, C.-C. Inflammatory response of macrophages in infection. *Hepatobiliary Pancreat. Dis. Int.* **2014**, *13*, 138–152. [CrossRef]
35. Li, S.; Gao, A.; Dong, S.; Chen, Y.; Sun, S.; Lei, Z.; Zhang, Z. Purification, antitumor and immunomodulatory activity of polysaccharides from soybean residue fermented with *Morchella esculenta*. *Int. J. Biol. Macromol.* **2017**, *96*, 26–34. [CrossRef] [PubMed]
36. Ren, Y.; Zheng, G.; You, L.; Wen, L.; Li, C.; Fu, X.; Zhou, L. Structural characterization and macrophage immunomodulatory activity of a polysaccharide isolated from *Gracilaria lemaneiformis*. *J Funct Foods* **2017**, *33*, 286–296. [CrossRef]

37. Thambiraj, S.R.; Phillips, M.; Koyyalamudi, S.R.; Reddy, N. Yellow lupin (*Lupinus luteus* L.) polysaccharides: Antioxidant, immunomodulatory and prebiotic activities and their structural characterisation. *Food Chem.* **2018**, *267*, 319–328. [CrossRef]
38. Park, S.B.; Song, H.M.; Kim, H.N.; Park, G.H.; Son, H.-J.; Um, Y.; Park, J.; Jeong, J.B. Anti-inflammatory effect of Biji (Soybean curd residue) on LPS-stimulated RAW264. 7 cells. *Korean J. Plant Resour.* **2018**, *31*, 117–123.
39. Lu, Y.; Li, W.; Yang, X. Soluble soybean polysaccharides enhance the protective effects of genistein against hepatic injury in high L-carnitine-fed mice. *Food Funct.* **2017**, *8*, 4364–4373. [CrossRef]
40. Meng, M.; Guo, M.; Feng, C.; Wang, R.; Cheng, D.; Wang, C. Water-soluble polysaccharides from *Grifola frondosa* fruiting bodies protect against immunosuppression in cyclophosphamide-induced mice via JAK2/STAT3/SOCS signal transduction pathways. *Food Funct.* **2019**, *10*, 4998–5007. [CrossRef]
41. Wang, X.; Chu, Q.; Jiang, X.; Yu, Y.; Wang, L.; Cui, Y.; Lu, J.; Teng, L.; Wang, D. *Sarcodon imbricatus* polysaccharides improve mouse hematopoietic function after cyclophosphamide-induced damage via G-CSF mediated JAK2/STAT3 pathway. *Cell Death Dis.* **2018**, *9*, 1–15. [CrossRef] [PubMed]
42. Zhou, D.; Huang, C.; Lin, Z.; Zhan, S.; Kong, L.; Fang, C.; Li, J. Macrophage polarization and function with emphasis on the evolving roles of coordinated regulation of cellular signaling pathways. *Cell. Signal.* **2014**, *26*, 192–197. [CrossRef] [PubMed]
43. Huang, W.-L.; Yeh, H.-H.; Lin, C.-C.; Lai, W.-W.; Chang, J.-Y.; Chang, W.-T.; Su, W.-C. Signal transducer and activator of transcription 3 activation up-regulates interleukin-6 autocrine production: A biochemical and genetic study of established cancer cell lines and clinical isolated human cancer cells. *Mol. Cancer* **2010**, *9*, 1–16. [CrossRef] [PubMed]

foods

MDPI

Article

Environmentally Friendly Hydrothermal Processing of Melon by-Products for the Recovery of Bioactive Pectic-Oligosaccharides

Xiana Rico, Beatriz Gullón and Remedios Yáñez *

Department of Chemical Engineering, Faculty of Science, University of Vigo (Campus Ourense), As Lagoas, 32004 Ourense, Spain; xianarico@protonmail.com (X.R.); bgullon@uvigo.es (B.G.)

* Correspondence: reme@uvigo.es

Received: 30 October 2020; Accepted: 18 November 2020; Published: 20 November 2020

Abstract: Melon by-products, that currently lack high value-added applications, could be a sustainable source of bioactive compounds such as polysaccharides and antioxidants. In this work, melon peels were extracted with water to remove free sugars, and the water-insoluble solids (WISs) were subjected to hydrothermal processing. The effect of temperature on the composition of the obtained liquors and their total phenolic content was evaluated. The selected liquors were also characterized by matrix assisted laser desorption/ionization-time of flight mass spectroscopy (MALDI-TOF MS), fourier transform infrared spectroscopy (FTIR) and high performance anion exchange chromatography with pulsed amperometric detection (HPAEC–PAD), and its phenolic compounds were identified and quantified by high-performance liquid chromatography-diode array detector-tandem mass spectrometry (HPLC–DAD–MS/MS). In addition, the spent solids from the hydrothermal treatment were characterized and their potential use was assessed. At the optimal conditions of 140 °C (severity 2.03), the total oligosaccharide yield accounted for 15.24 g/100 g WIS, of which 10.07 g/100 g WIS were oligogalacturonides. The structural characterization confirmed the presence of partially methyl esterified oligogalacturonides with a wide range of polymerization degrees. After precipitation, 16.59 g/100 g WIS of pectin were recovered, with a galacturonic acid content of 55.41% and high linearity.

Keywords: melon by-products; hydrothermal treatment; pectin-derived oligosaccharides; oligogalacturonides; phenolic compounds; antioxidant activities

1. Introduction

Around 1300 million tonnes of food are discarded every year, between production and consumption [1]. For instance, in Europe alone, 88 million tonnes of food waste were generated in 2012, of which about 39% arises from manufacturing activities [2,3]. It would be equivalent to 20% (*w/w*) of the total food produced and a generation of 173 kg of food waste per person per year [3].

In particular, waste and by-products from the fruit and vegetable processing industries, which are characterized by a high fraction of discards, are in fifth place in the ranking of total food waste in Europe, assuming 8% of the total [4,5]. These wastes, with high contents of moisture and microbial loads, have an important contribution to the global environmental burden, and currently, they are mainly destined for low value-added applications, such as composting or animal feed [4,5]. In addition, the requirements of European regulations regarding the prevention, management and landfill of waste (Directives (EU) 2018/851 and (EU) 2018/850), support the growing need to find alternatives for their valorization that allow them to be re-incorporated into productive processes as raw materials, thus adopting the principles of circular economy and the concept of industrial symbiosis [6,7].

In this context, one of the most consumed fruit crops worldwide is melon (*Cucumis melo* L.) [8]. According to information provided by the Food and Agriculture Organization of the United

Nations (FAO), its annual production (*Cucumis melo* L.) has recorded slight fluctuations over the last ten years, between 25.5 and 27.3 million tonnes [9]. It is often used for minimal fresh processing as cylinders, cubes or slices [10] or for the production of juices, jams, compotes or dehydrated pulp [11,12]. The processing by-products account for around one third of the weight of the fruit and would be mostly made of skin and lower percentages of pulp and seeds [13].

Melon by-products have high contents of polysaccharides and minerals, as well as bioactive compounds with antioxidant activity, including phenolic compounds such as flavonoids and phenolic acids, with important applications in several food and pharmaceutical industries [14–18]. One of the main components of the melon peel is pectin [19,20], a complex heteropolysaccharide found in the plant cell walls of many agro-industrial by-products, consisting of a backbone of galacturonic acid (GalA) with alternating 'smooth' and 'hairy' regions [21,22]. Homogalacturonans (HGs) represent the smooth regions, which are composed of GalA linked by α-(1,4) glycosidic bonds and constitute about 65% of the pectin, while rhamnogalacturonans (RG-I and RG-II) form the hairy region, with RG-I being the main branched structure of pectin, constituting 20–35% of the molecule [21]. In RG-I, the GalA backbone is often interrupted by rhamnose units bearing neutral sugar side chains composed mainly of galactose and arabinose [22]. This polymer displays important applications in the food industry as thickener, gelling agent, stabilizer, encapsulant and edible food packaging films [23–25]. Moreover, pectin-derived oligosaccharides are reported as emerging prebiotics and in recent years they are being studied for their potential health benefits [21,23,26].

In the last decade, autohydrolysis has been applied successfully for the extraction of pectic oligosaccharides from several fruit by-products, such as lemon peels [27], orange peels [28], sugar beet pulp [29], pomegranate peels [30] or mango peels [31]. The hydrothermal processing, carried out in pressurized hot water, is catalyzed by the hydronium ions of the reaction media as well as by the acids generated in situ. This green technology allows the selective solubilization of hemicelluloses, yielding a pre-treated solid fraction enriched in cellulose and lignin, suitable for further applications in a bio-refinery context [32–35]. Temperature and time are the main factors affecting this treatment, and their effect can be measured by combining the two variables into a single parameter called the severity factor (R_0) [33].

The application of bioactive compounds in the formulation of novel functional foods has been drawing the attention of researchers due to the health benefits associated with their consumption and the high demand and acceptance that they nowadays are having in the market [36–38]. Melon by-products, that currently lack high value-added applications, could be considered as a cheap and sustainable source of these compounds [5,39]. However, the literature concerning extraction technologies for the production of bioactive compounds from melon by-products is limited. For instance, the extraction of melon peel pectin with citric acid has been optimized recently by Muthukumaran et al. [20]. In another study, Raji et al. [19] assessed the influence of several acids on the extraction of melon peel pectin, obtaining the highest pectin yield (29%) with citric acid. Several studies focused on the recovery of the phenolic fraction by conventional alcoholic extraction [8,14,40], aqueous wet grinding [41] or ultrasound assisted extraction [42] have also been reported. More recent studies [11,43] evaluated the extraction of bioactive compounds (total phenolics, chlorophylls, total carotenoids and vitamin C) from several melon parts by ultra-turrax homogenization in alcoholic media.

However, to the best of our knowledge, despite the important content of pectin of the melon peels, no literature dealing with the extraction and characterization of pectin-derived oligosaccharides with phenolic content from these by-products is available. Hence, the need for additional research focused on its extraction and characterization, since the polysaccharide structure has an important role in their functionality and applications [23,44].

Therefore, this work evaluates the suitability of non-isothermal autohydrolysis for the production of melon by-products extracts enriched in functional pectic oligosaccharides with antioxidant activity. The liquid phases obtained were assayed for their chemical composition, total phenolic contents and antioxidant activities (by DPPH (α,α-diphenyl-β-picrylhydrazyl),

ABTS (2,2-azino-bis-3-ethylbenzothiazoline-6-sulphonic acid) and FRAP (ferric reducing antioxidant power) methods). The identification and quantification of the main phenolic compounds was also performed by HPLC–DAD–MS/MS. Additional information about the pectic oligosaccharides composition has also been provided by FTIR and MALDI-TOF. Finally, in order to propose further bio-refinery stages, the spent solids were characterized and their susceptibility to enzymatic hydrolysis was also studied.

2. Materials and Methods

2.1. Standards

The standards used for determination were: arabinose (99%), xylose (99%), rhamnose monohydrate (99%), galacturonic acid monohydrate (97%), acetic acid (96%), formic acid (99%), gallic acid (97.5%), trolox (97%), digalacturonic acid (85%), trigalacturonic acid (90%) and polygalacturonic acid (90%), purchased from Sigma-Aldrich (Steinheim, Germany); glucose (98%), purchased from Scharlau (Barcelona, Spain); galactose (99%), mannose (99%) and fructose (98%), purchased from Panreac (Barcelona, Spain). For the identification of phenolic compounds, 4-hydroxybenzoic acid (99%), p-coumaric acid (98%), gallic acid (97.5%), salicylic acid (99%), tyrosol (98%), sinapic acid (98%) and caffeic acid (98%) were purchased from Sigma-Aldrich (Steinheim, Germany); ferulic acid (99%) and vanillic acid (97%) were purchased from Sigma-Aldrich-Fluka (Steinheim, Germany).

2.2. Raw Material

Melon by-products (var. piel de sapo) consisting of peels with a small amount of seeds were kindly supplied by FreshCut, S.L. (Vigo, Pontevedra, Spain), a company dedicated to the development and marketing of fresh cut products. They were cut into small pieces and frozen at −18 °C until use.

2.3. Aqueous Extraction of Melon by-Products

Melon by-products (MPs) were centrifuged (SV4028, AEG Electrolux) at 2800 rpm for 15 min, separating the extracts from the solid fraction. The solid fraction was then washed with distilled water for 15 min at room temperature in a 2 L stirred reactor (20 g of water/g of oven dry MP), and centrifuged again for 15 min. The resulting water insoluble solids (WISs) were ground in a regular coffee grinder and then frozen until use, except a small fraction that was dried in an oven at 60 °C for analysis.

2.4. Autohydrolysis

Wet WIS samples were mixed with water in a liquid to solid ratio of 20 kg/kg (oven dry basis) in a 0.6 L stainless steel reactor (model 4842 from Parr Instruments, Moline, IL, USA). Based on the literature concerning pectin-rich raw materials [27–29] and on preliminary experiments (data not shown), the hydrothermal treatments were performed under non-isothermal conditions in the range of temperatures from 130 to 165 °C. Figure 1 displays the temperature profiles followed during the heating and cooling of all the experiments carried out in this work. With the purpose of facilitating the comparison between different reactors, the combined effects of temperature and time can be expressed in terms of the severity (S_0). This parameter is defined as the logarithm of the severity factor R_0 [45] and it is calculated by the following equation:

$$\begin{aligned} S_0 = \log R_0 &= \log(R_{0HEATING} + R_{0COOLING}) \\ &= \log\left(\int_0^{t_{Max}} \exp\left(\frac{T(t)-T_{REF}}{\omega}\right) \times dt + \int_{t_{Max}}^{t_F} \exp\left(\frac{T'(t)-T_{REF}}{\omega}\right) \times dt\right) \end{aligned} \quad (1)$$

where t_{Max} is the time (min) needed to achieve the target temperature of each treatment (t_{Max}, °C); t_F is the time (min) of cooling; T(t) and T′(t) (°C) represent the temperature profiles in the heating and cooling stages, respectively; and ω and T_{REF} are parameters whose values are fixed according to the literature (ω = 14.75 °C; T_{REF} = 100 °C). The calculated values of S_0 for the performed experiments are also included in Figure 1.

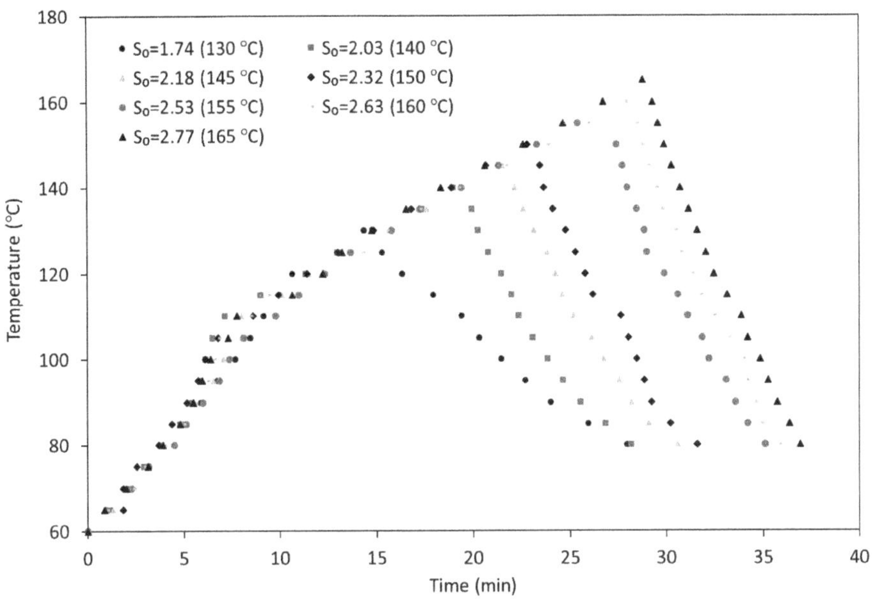

Figure 1. Heating and cooling temperature profiles determined for the autohydrolysis experiments carried out at all conditions assayed.

After each autohydrolysis treatment, the solid and liquid phases were separated by manual pressing using a cloth bag. The liquid phases were weighed to determine liquor recovery and stored at 4 °C until further analysis, and the solid phases were washed with water, dried in an oven at 60 °C, weighed and subjected to moisture determination in order to calculate the solid yield. Aliquots of the autohydrolysis liquors and the spent solids of the optimal treatment were analyzed using the methodology described below.

2.5. Pectin Precipitation

Ethanol was added to the autohydrolysis liquor obtained at the optimum conditions to a final concentration of 75% *v*/*v*. After mixing, the solution was allowed to stand at 4 °C overnight, after which it was centrifuged (Rotixa 50 RS centrifuge, Hettich Lab Technology, Tuttlingen, Germany) at 4000 rpm (radius of 211 mm) for 20 min and subsequently washed with 96% ethanol. After separation by centrifugation, the washed precipitate was oven dried at 55 °C until constant weight and redissolved in water for analysis.

2.6. Enzymatic Hydrolysis of Autohydrolyzed Solids

The dried spent solids from hydrothermal treatment at the selected conditions were treated by enzymatic hydrolysis to assess their digestibility. The enzyme cocktail used contained endopolygalacturonase (Viscozyme L from *Aspergillus aculeatus*, 50 U/g spent solid), cellulase (Celluclast 1.5 L, 12 filter paper unit (FPU)/g spent solid) and β-glucosidase (Novozym 188, 5 international unit (IU)/FPU cellulase). The enzymes used in this study were kindly supplied by Novozymes, Madrid, Spain. The hydrolysis was carried out at 37 °C and pH = 5 for 48 h.

2.7. *Analytical Methods*

2.7.1. Analysis of the WISs and Spent Solids from Autohydrolysis Treatment

The dried WISs and solids from hydrothermal treatment were subjected to moisture (TAPPI T-264-om-88 method) and ash (T-244-om-93 method) determination. Protein was calculated using the nitrogen content (6.25 g protein/g nitrogen), which was determined with a Thermo Fisher Quest Flash EA 1112 analyzer, using 130 and 100 mL/min He and O_2 and an oven temperature of 50 °C. Metals were analyzed using a fast sequential atomic absorption spectrometer after digestion in an MLS-1200 Microwave Labstation mega with 5 mL of HNO_3 65% (*w/w*), 1 mL of H_2O_2 30% (*w/v*) and 0.5 mL of HF 40% (*w/w*). Conventional quantitative acid hydrolysis with 72% (*w/w*) H_2SO_4 (TAPPI T13 m method) was used to determine the content of hemicelluloses, glucan and lignin. The solid residue from hydrolysis was recovered by filtration, oven-dried and considered as Klason lignin (TAPPI T13 m assay), and the liquid phase was analyzed for monosaccharides (rhamnose and arabinose) and acetic acid (coming from acetyl groups) by high-performance liquid chromatography (HPLC) using an Agilent 1260 equipped with a refractive index (RI) detector with an Aminex HPX-87H column (BioRad, Life Science Group, Hercules, CA, USA) operating as follows: mobile phase, 3 mM H_2SO_4; flow, 0.6 mL/min; temperature, 50 °C. Another separation with an Aminex HPX-87P column (BioRad, Life Science Group, Hercules, CA, USA) was performed for the determination of glucose, xylose, galactose, arabinose and mannose, operating with deionized water as the mobile phase, at a flow rate of 0.4 mL/min at 80 °C. The arabinose content was calculated as the average of the results obtained by both columns. The liquid phase was also subjected to uronic acid determination by spectrophotometry using galacturonic acid as a standard for quantification [46].

2.7.2. Chemical Characterization of Liquors

The samples of liquors from hydrothermal treatments were filtered through 0.45 μm membranes and used for the direct HPLC determination of galacturonic acid, rhamnose, arabinose, formic acid and acetic acid using the same method employed in the analysis of the WIS fraction with an Aminex HPX-87H column. Another aliquot of filtered liquors was quantified for glucose, xylose, galactose, arabinose, mannose and fructose using an Aminex HPX-87P as described for the WIS fraction. For the determination of OGalA (oligogalacturonides), another sample of liquors was subjected to total enzymatic posthydrolysis using cellulases (Celluclast 1.5 L) and an endopolygalacturonase (Viscozyme L from *Aspergillus aculeatus*) at 37 °C and pH = 5 for 40 h, with a cellulase loading of 5 FPU/g liquor and an endopolygalacturonase loading of 45 U/g liquor [27,47]. The neutral oligosaccharides were quantified by quantitative posthydrolysis (4% (*w/w*) sulphuric acid at 121 °C for 20 min). The reaction products of both posthydrolyses were assayed by the same HPLC methods using the already mentioned columns. The increase in the concentrations of monosaccharides and acetic acid caused by posthydrolysis provided a measure of the oligomer concentration and their degree of substitution by acetyl groups (Ac-O). OSs (oligosaccharides) were expressed as monosaccharide equivalents. The content of non-volatile compounds (NVCs) was measured by oven-drying at 60 °C until constant weight. Other non-volatile compounds (ONVCs) were calculated by difference as (NVCs−monosaccharides−oligosaccharides)/NVC. Protein and ash were determined using the same methods as for the solid fractions. All determinations were made in triplicate.

2.7.3. Total Phenolic Content (TPC)

The total phenolic content of autohydrolysis liquors was determined according to the Folin–Ciocalteau method [48]. Aliquots of the samples (500 μL), water as blank and gallic acid standards (10–80 mg/L) were mixed with water (3.75 mL), and Folin–Ciocalteu chemical (1:2 *v/v*; 250 μL) and Na_2CO_3 (10% *w/v*; 500 μL). These mixtures were kept in darkness at room temperature for 1 h and their absorbance was then measured at 765 nm. The results were expressed as mg of gallic acid equivalents (GAE)/g dried WISs. All samples were analyzed in triplicate.

2.7.4. Antioxidant Activity

Antioxidant capacity was evaluated using three different methods, namely DPPH (α,α-diphenyl-β-picrylhydrazyl), ABTS (2,2-azino-bis-3-ethylbenzothiazoline-6-sulphonic acid) and FRAP (ferric reducing antioxidant power) according to the methodology described by Gullón et al. [49]. In all antioxidant capacity assays, Trolox was used as a standard and results were expressed as mg Trolox equivalents (TE)/g dried WISs as the mean of three replicates.

2.7.5. Identification and Quantification of Major Phenolic Compounds Using HPLC–DAD–MS/MS

The profile of phenolic compounds was analyzed using an Agilent model 1260 Infinity (Palo Alto, CA, USA) connected directly to a mass detector AB SCIEX Triple Quad 3500 (AB Sciex, Foster City, CA, USA) equipped with a turbo V™ electrospray ionization source (ESI), and to a Waters 996 photodiode array detector. The separation was carried out on a Phenomenex Luna C18 column (150 mm × 2 mm; 3 μm), with an injection volume of 5 μL and a flow of 300 μL/min. The mobile phase was 0.1% (*v*/*v*) formic acid in water (A) and 0.1% (*v*/*v*) formic acid in acetonitrile (B). The gradient used was: 98% A (*v*/*v*), 0–4 min; 98–80% A (*v*/*v*), 4–7 min; 80–10% A (*v/v*), 7–14 min; 10% A (*v/v*), 14–15 min; 10–98% A (*v/v*), 15–17 min. The mass spectrometer was run in positive and negative ionization, using N_2 as the nebulizer and collision gas, with an ion spray voltage of 4500 V, source temperature of 400 °C and nebulizer gas pressure of 55 psi. The compounds were identified by the comparison with the corresponding standards, based on retention time data, extracted ion chromatograms and MS/MS spectra. For the quantification, calibration curves were made by using different concentrations of the standards and plotting them against peak areas.

2.7.6. Structural Characterization of the Extracted Oligogalacturonides (OGalA)

To obtain detailed information on the structural characteristics of solubilized oligogalacturonides at the optimal temperature of autohydrolysis, the liquors were freeze-dried and were analyzed using different analytical techniques including FTIR, HPAEC-PAD and MALDI-TOF.

Fourier Transform Infrared Spectroscopy (FTIR)

The FTIR analysis of the oligogalacturonides was performed on a Nicolet 6700 Spectrometer. A total of 34 scans were accumulated in transmission mode with a resolution of 4 cm^{-1}. The spectrum was obtained in a range of 4000–400 cm^{-1}.

High Performance Anion Exchange Chromatography with Pulsed Amperometric Detection (HPAEC–PAD)

The oligogalacturonides were analyzed by HPAEC–PAD using an ICS3000 chromatographic system (Dionex, Sunnyvale, CA, USA), fitted with a CarboPac PA-1 column (2 mm i.d. × 250 mm) in combination with a CarboPac PA guard column (2 mm i.d. × 25 mm) and an ISC3000 PAD detector. HPAEC–PAD was performed following the pectic oligosaccharides (POS) identification method described by Gómez et al. [27]. GalA with DP 1, 2 and 3 were used as standards, as well as a mixture of OGalA prepared by the hydrolysis of a 1% (*w/w*) solution of commercial polygalacturonic acid at 121 °C for 40 min at pH = 4.4 adjusted with NaOH [27].

Matrix Assisted Laser Desorption/Ionization-Time of Flight Mass Spectroscopy (MALDI-TOF MS)

The absolute masses of oligogalacturonides were obtained by MALDI-TOF MS, using an Ultraflex workstation (Bruker Daltonics, Billerica, MA, USA), operating in reflectron mode and positive polarity. Sample preparation was performed according to the protocol described by Gómez et al. [50] using 2,5-dihydroxybenzoic acid (DHB) as the matrix. Data were acquired and processed by means of the Flex Control and Flex Analysis software (Bruker Daltonics, Billerica, MA, USA), respectively.

3. Results and Discussion

3.1. Chemical Characterization of the Raw Material

As is the case for other food by-products, the MPs are characterized by a high moisture content (90.05%). In order to eliminate free sugars and other water-soluble compounds, which would decompose into undesirable compounds during the hydrothermal treatment, a centrifugation stage and aqueous extraction followed by further centrifugation was proposed. The chemical composition of the MP extracts and water insoluble solid (WIS) fractions obtained is shown in Table 1. The extractives accounted for 42.21% of the raw material and contained mostly sugars (especially glucose and fructose) and some protein and organic acids. The remaining WISs (57.79% of the raw material) were mainly made up of glucan (includes glucose from cellulose and other glucose moieties) and Klason lignin (24.54 and 19.96%, respectively), followed by protein and galacturonan (11.36 and 11.99%), and minor amounts of other polysaccharides or substituents. Ash and some minerals were also determined, highlighting the content of potassium and magnesium. In general, the presented data are in range of that found in recent literature for melon peels [8,15,16,51]. However, the results showed sample variability and reported clear differences in the chemical composition, for instance, in the protein and ash contents (ranges from 3.25 to 19.14% and from 3.67 to 11.13%, respectively). They are thought to be related with the variety of plant used, the state of ripening of the fruit and/or the analytical methodology used in each case.

Table 1. Chemical composition of the aqueous extracts and the water-insoluble solids (WISs) on a dry basis.

Component	Content
Extracts (g/100 g extracts)	
Glucose	36.40 ± 3.18
Sucrose	5.35 ± 0.24
Fructose	45.79 ± 0.26
Uronic Acids	2.79 ± 0.04
Citric Acid	5.59 ± 1.00
TPC	0.89 ± 0.02
Protein	10.81 ± 1.05
WIS (g/100 g WIS)	
Glucan	24.54 ± 0.10
Xylan	4.89 ± 0.21
Galactan	3.31 ± 0.21
Mannan	1.54 ± 0.04
Arabinosyl S.	1.56 ± 0.11
Acetyl Groups	2.05 ± 0.35
Galacturonan	11.99 ± 0.69
Klason lignin	19.96 ± 1.72
Protein	11.36 ± 1.86
Ash	3.36 ± 0.03
mg/100 g WIS	
Iron	8.74 ± 3.97
Potassium	863.01 ± 109.81
Manganese	1.83 ± 0.36
Magnesium	218.73 ± 50.57

Compared with other pectin-rich raw materials that have also been used for pectic oligosaccharides extraction, such as lemon and orange peels, MPs contain a similar glucan content, considerably higher proportions of lignin and protein as well as a lower amount of galacturonan [27,28]. However, the sugar beet pulp presented comparable protein content but a much higher percentage of arabinosyl substituents and lower galacturonan and lignin [29].

3.2. Effect of Hydrothermal Processing on Autohydrolysis Liquors

WIS samples were subjected to non-isothermal autohydrolysis at maximum temperatures in the range of 130–165 °C or S_0 between 1.74 and 2.77, according to temperature profiles shown in Figure 1. This treatment mainly causes the solubilization of the pectin, protein and other hemicellulosic fractions as well as low molecular weight phenolic compounds, yielding spent solids with increased proportions of glucan and lignin [27–30]. Therefore, as expected, in this study, the percentage of solubilization of WISs increased with the temperature of the treatment reaching a maximum value of 36.46 g/100 g WIS in the experiment performed at 165 °C. A similar trend has been previously observed during the autohydrolysis treatment of lemon and oranges peels, reaching higher percentages of solubilization in both cases, with values close to 50% and greater than 60%, respectively, under the same operational conditions [27,28].

Depending on the severity of the treatment, the autohydrolysis liquors contain different proportions of high molecular weight polymers, oligomers as well as monomers and sugar degradation products, among others. The degree of polymerization and the substitution pattern of the extracted oligosaccharides are also affected by the treatment severity. It is well known that the structural attributes play an important role, since it is evident that the chemical and physicochemical features can affect, for instance, the cholesterol-lowering ability, prebiotic potential or emulsifying properties [21,23]. Therefore, the successful modulation of the treatment severity can be used to obtain tailored oligosaccharides suitable for industrial/food applications and to limit carbohydrate decomposition or lignin solubilization, which would result in the presence of undesired compounds in liquors.

3.2.1. Liquor Composition

Most of the autohydrolysis soluble products are non-volatile compounds (NVCs), which are quantifiable by oven-drying the liquors. These NVCs are made of oligosaccharides, monosaccharides, organic acids and other compounds (ONVCs). This fraction includes protein, protein-derived compounds, soluble minerals, and acid soluble lignin.

Figure 2 shows the severity dependence of the mass fraction of NVCs and ONVCs in the autohydrolysis liquors (expressed as g per 100 g WIS treated). The lowest NVC content was noted at 130 °C, and then this parameter gently increased in experiments carried out at temperatures between 140 to 165 °C, with values between 23.22 and 28.99%. Regarding ONVCs, in this kind of raw materials when operating at low to moderate severities, this fraction would contain mostly protein and soluble minerals. However, as can be seen in Figure 2 when the autohydrolysis was performed at 165 °C, higher NVC and ONVC contents have been detected in the reaction media. This fact would be related to the increase in the solubilization of WISs and to the presence of decomposition reactions, which would result in the generation of undesirable compounds, such as furans or organic acids, also reported in previous studies [27,52]. Oligosaccharides made the most part of the NVC at all the treatment temperatures, fluctuating between 59 and 66% of this fraction at temperatures of 130–155 °C and then decreasing down to 54% at 165 °C, due to further depolymerization and degradation reactions. In addition, other components present in the NVC of all the autohydrolysis liquors are monosaccharides (in the range of 8–10 g per 100 g NVC) and organic acids (6–11 g per 100 g NVC).

The ratio of total OSs to ONVCs increased from 2.5 to 3.5 at temperatures from 130 to 140 °C and then gradually decreased down to 2.2 at 165 °C, similarly to the results reported for lemon and orange peels and sugar beet pulp, where this ratio decreased from 4 or 5 at the lowest temperatures assayed to 1 or 0 at the most severe ones [27–29].

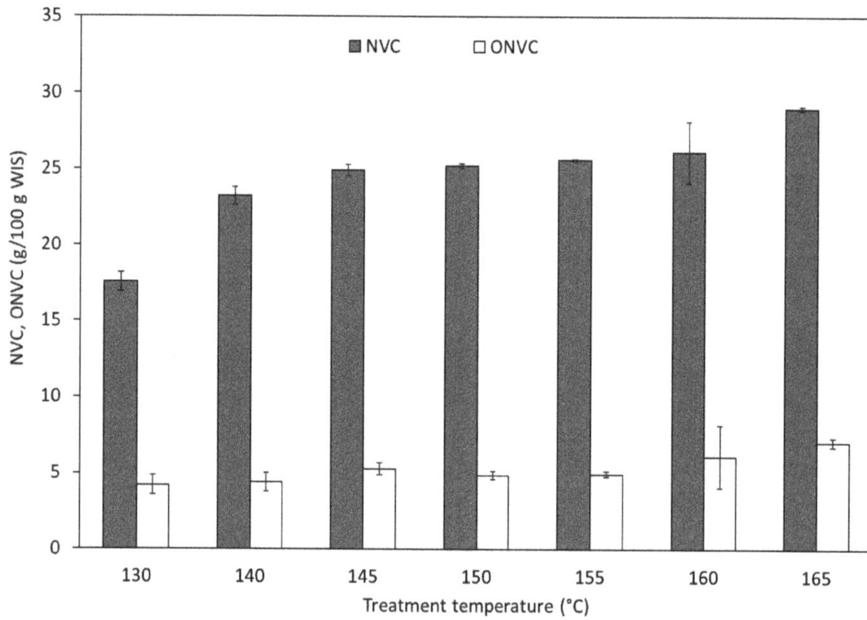

Figure 2. Temperature dependence of the concentration of NVCs and ONVCs in the reaction liquors. NVC, non-volatile compound; ONVC, other non-volatile compound or impurity.

In order to provide additional information about the chemical composition of autohydrolysis liquors, Figure 3 shows quantitative information on oligosaccharides (determined as monosaccharide equivalents and expressed as g per 100 g WIS treated). The total oligosaccharides content increased with the severity of the treatment up to reach a maximum of 16.28 g/100 g WIS in the experiment performed at 150 °C. As can be observed in Figure 3, oligogalacturonides (OGalA), the main component of the liquors, showed a similar pattern, with the highest contents being between 140 and 150 °C. They exhibited a maximum of 10.41 g/100 g WIS at 150 °C, corresponding to a conversion yield of 80.02% (g of OGalA monomer equivalents in liquors/100 g of galacturonan monomer equivalents in the WISs). As expected, higher severities led to liquors with decreased OGalA suggesting their depolymerization. Other oligosaccharides detected were mostly GalOS (galactose units in oligosaccharides), followed by AraOS (arabinose units in oligosaccharides), and smaller contents of glucose, acetic acid, mannose, rhamnose and xylose units in oligosaccharides (GOS, Ac-OS, ManOS, RhaOS and XOS, respectively). The temperature dependence of GalOS, AraOS, RhaOS and Ac-OS is shown in Figure 3, whereas the sum of ManOS, XOS and GOS by simplicity has been included in the other oligosaccharides fraction. In general terms, the percentages of GalOS, AraOS, RhaOS and Ac-OS gradually increased in all the temperature range evaluated, reaching values of 3.87, 1.53, 0.55 and 0.61 g/100 g WIS, respectively, at 165 °C. However, other compounds such as GOS have experienced only slight variations, remaining close to an average value of 0.56 g/100 g WIS in all the conditions assayed.

In this study, galacturonan presented higher susceptibility to hydrolysis reactions than other polymers which required greater treatment severity. For instance, Galacturonan into OGalA, Galactan into GalOS and Arabinosyl S. into AraOS conversions close to 60% were obtained in experiments carried out at 130, 150, and 140 °C, respectively. A maximum temperature of 165 °C was required to reach a complete conversion of Galactan into GalOS, whereas in the case of OGalA the maximum was obtained at 150 °C (80.02%). Similar patterns have been found for lemon and orange peels and sugar beet pulp, with the highest OGalA solubilization at temperatures between 150 and 160 °C, whereas higher severities were needed for the optimum extraction of other oligosaccharides [27–29].

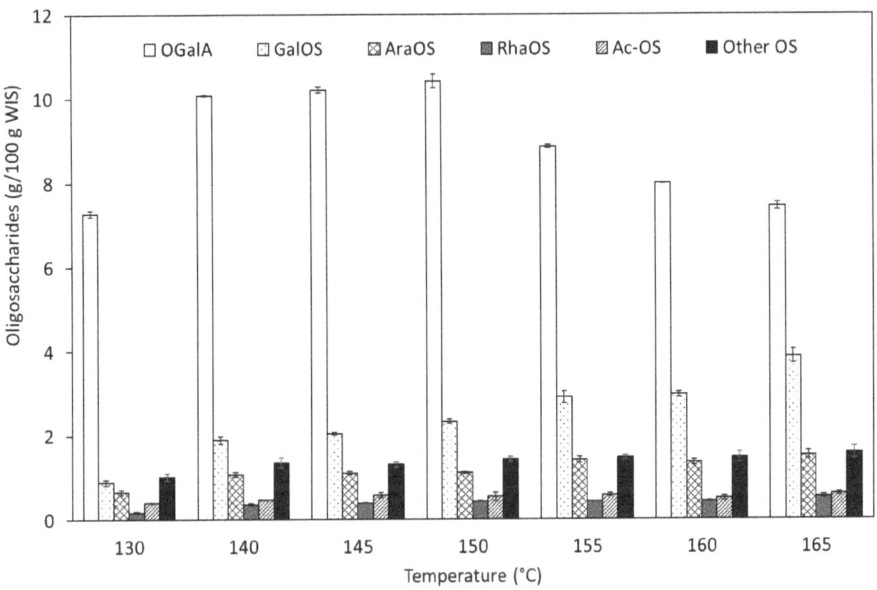

Figure 3. Temperature dependence of the oligosaccharides. OGalA: oligogalacturonides; GalOS: galactose units in oligosaccharides; AraOS: arabinose units in oligosaccharides; RhaOS: rhamnose units in oligosaccharides; Ac-OS: acetic acid units in oligosaccharides; OS: oligosaccharides.

The selectivity of the autohydrolysis treatment was confirmed by the low conversion yield of glucan into GOS which oscillated around 2% for all the severities assayed, therefore leaving solid fractions with interesting proportions of glucan and applications in further biorefinery stages. Other studies working with higher maximum temperatures found an increasing GOS yield with increasing temperature [27–29]. The total oligosaccharides to monosaccharides ratio, with values between 6 and 9 (with the maximum at 150 °C), also gives insight into the selectivity of the process. The monosaccharides (MS) fraction showed small variations in the interval studied, with an average value close to 2 g/100 g WIS. It was mainly constituted by fructose and glucose, followed by galactose and arabinose (coming from arabinan, galactan or arabinogalactan) and some xylose. There is a lot of variation in the OS/MS ratios found in the literature for liquors rich in pectic oligosaccharides recovered by autohydrolysis. For instance, lemon peel liquors had ratios decreasing from 6 to 2 when increasing the treatment temperature [27], while orange peel and sugar beet pulp liquors had much higher ratios with maximums of 16 and 47, respectively, both at a treatment temperature of 150 °C [28,29]. Formic acid was the main degradation product generated from glucose and fructose degradation, whose production increases with the treatment temperature up to 0.72 g/100 g WIS at 165 °C. This value is considerably higher than those found at the same temperature in sugar beet pulp autohydrolysis liquors (0.38 g/100 g sugar beet pulp) [29], lemon peel waste liquors (about 0.16 g/100 g WIS) [27], and orange peel liquors (about 0.21 g/100 g WIS) [28].

The amount of liquors recovered was also influenced by the treatment severity. At 130 °C only a 75% (*w*/*w*) of the liquors were recovered, lowering the yields obtained. At temperatures from 140 to 160 °C this parameter reached a value of about 85% and lastly at 165 °C it increased to 90%. Previous studies with similar raw materials related higher water retention capacity with the pectin content of the solids [27,28], therefore, increased liquor recovery would be associated with increased pectin fraction solubilization.

Taking into account that there are no important differences in the results obtained for the recovery of OGalA in the temperature range between 140 and 150 °C, the lowest temperature, 140 °C, was chosen as the optimum treatment temperature, resulting in an average OGalA content of 10.07 g/100 g WIS, corresponding to a yield of 77.42% and a total oligosaccharide content of 15.24 g/100 g WIS. In this study, a lower operating temperature allowed higher yields than the ones achieved in the optimum for water insoluble substrates of orange peels (66.23% at 160 °C) and sugar beet pulp (56.77% at 160 °C) and similar to that in water insoluble substrates of lemon peels (79.32% at 160 °C). However, due to a minor presence of galacturonan in the raw material, the resulted OGalA content was lower for MPs than for lemon, orange and passion fruit peels (18.28, 13.96 and 13.78 g/100 g WIS), sugar beet pulp, two different cultivars of pomegranate peels and pomelo peels (12.06, 13.58, 14.71, 15.04 g/100 g raw material, respectively) [27–30,53,54]. There have also been studies that yielded a lower OGalA content, probably due to a lower pectin content in the raw material: 8.46 and 6.49 g/100 g raw material for apple pomace and cacao pod husks and 7.82 g/100 g WIS for jackfruit peels [55–57]. Under the selected operational conditions, the degree of acetylation was 15.26 mol of acetic acid/100 mol of galacturonic acid, in between those determined in the optimum for orange peels and sugar beet pulp (5.56 and 50.66 mol of acetic acid/100 mol of galacturonic acid, respectively) [28,29].

The non-volatile impurities were 18.97 g/100 g NVC, in range with those reported for lemon and orange peels and sugar beet pulp (15–18 g/100 g NVC) [27–29]. However, further analysis of the liquors allowed concluding that in MPs, the ONVCs would be justified by protein and ash, with contents of 2.82 g/100 g WIS (corresponding to a solubilization yield of 24.80 g protein in liquor/100 g protein in WISs) and 2.27 g/100 g WIS (solubilization yield of 67.34 g ash in liquor/100 g ash in WISs), respectively. Sugar beet pulp contains a similar amount of protein to water insoluble melon substrate. However, the generally lower protein yields in the recovered liquors (3.95 to 27.09 g protein in liquor/100 g protein in sugar beet pulp at temperatures from 140 to 180 °C) could explain the lower ONVCs reported previously for this raw material [29].

3.2.2. Antioxidant Potential

Another aspect assessed in this work was the antioxidant potential of the liquors from MPs. Phenolic compounds are commonly recovered from fruits and vegetables by aqueous–organic extractions, but an important part of the phenols contained in these raw materials are not extractable by conventional methods. They are called macromolecular antioxidants and include high molecular weight phenols and the low molecular weight ones linked to macromolecules like pectin [58,59]. Hence, there is interest in studying new technologies for the recovery of phenolic compounds in pectin-rich raw materials, since its solubilization could allow access to a macromolecular fraction and thereby increasing its extraction performance.

In this context, this study proposes the hydrothermal treatment, since it has been previously studied for the same purpose with several raw materials such as vine shoots, spent coffee grounds, peanut and hazelnut shells, pomegranate and mango peels and purple corn cob, among others [30,31,49,52,60–62]. According to the literature data, the recovery of phenolic compounds by this technology could be achieved by partial lignin depolymerization and by the hydrolysis of low molecular weight phenols bound to oligosaccharides [49,63].

With the aim of knowing the impact of hydrothermal treatment on WISs, the total phenolic content (TPC) and antioxidant activities (DPPH, FRAP and ABTS assays) of all the autohydrolysis extracts obtained has been determined. As can be seen in Figure 4, both TPC and activities increased with temperature, reaching the highest values at the highest temperature studied (TPC of 8.96 mg GAE/g WIS and activities of 3.19, 4.85 and 19.63 mg TE/g WIS determined by DPPH, FRAP and ABTS methods, respectively). Moreover, all the antioxidant activities assayed are in good correlation with the TPC, with R^2 of 0.89, 0.96 and 0.84 for DPPH, FRAP and ABTS, respectively.

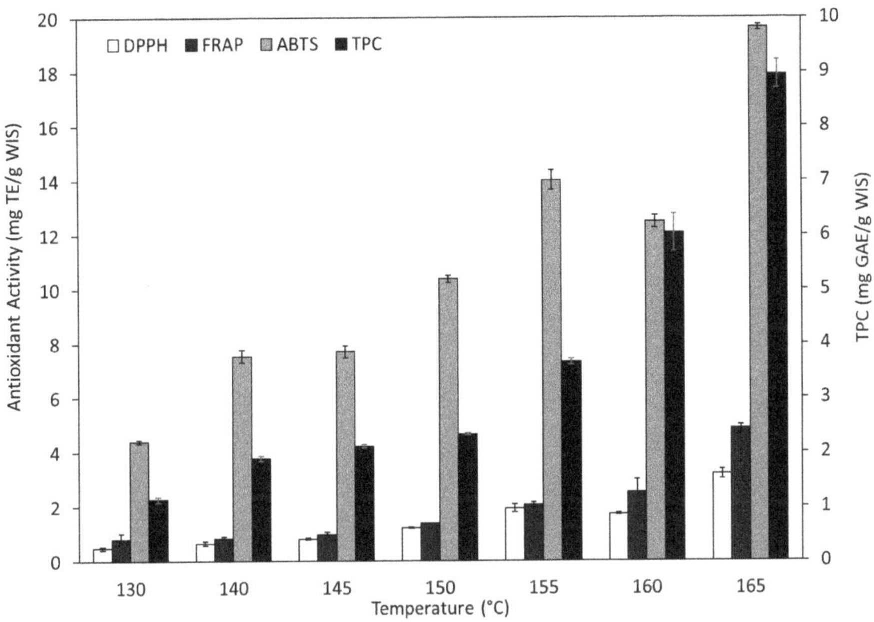

Figure 4. Effects of the autohydrolysis temperature on the total phenolic content (TPC) and antioxidant capacity of the extracted liquors by DPPH (α,α-diphenyl-β-picrylhydrazyl), ABTS (2,2-azino-bis-3-ethylbenzothiazoline-6-sulphonic acid) and FRAP (ferric reducing antioxidant power) assays.

According to Pérez-Jiménez and Saura-Calixto [58], the melon peel of the "Piel de sapo" variety presents a polyphenol content of 9.68 mg GAE/g dw (dry weight) peels, of which 67% are non-extractable. On this basis, the autohydrolysis treatment at 165 °C would have extracted 93% of the total amount contained in the raw material. Although it would be feasible to improve the extraction yield increasing treatment temperatures, with the reported information, it seems reasonable to assume that 165 °C could be close to the optimal treatment temperature for the recovery of phenols. However, due to the typical variability of this kind of raw materials, to confirm this assumption still requires a more exhaustive characterization.

At the optimal temperature selected for the production of pectic oligosaccharides, 140 °C, the phenolic content and antioxidant activities were much lower (TPC: 1.89 mg GAE/g WIS, DPPH: 0.65 mg TE/g WIS, FRAP: 0.83 mg TE/g WIS and ABTS: 7.52 mg TE/g WIS). The TPC was lower than the total extractable polyphenols (3.16 mg GAE/g MP) determined by Pérez-Jiménez and Saura-Calixto [58] with the same melon variety, and within the range provided for the peels of other melon varieties when the extraction was carried out using conventional solvents (1.11–7.03 mg GAE/g MP) [14,40,42]. Certain studies focused on the co-production of oligosaccharides and phenols in aqueous media from peanut shells, spent coffee grounds, chestnut shells, or purple corn cob resulted in higher TPC contents; but in all these cases, given the different nature of the substrates, a considerably higher treatment severity was required to reach the optimal extraction efficiency [52,62,64,65]. The recovery of the phenolic compounds that remain in the solid fraction after melon peel treatment at 140 °C is key for the complete valorization of melon by-products. With this aim, a sequential second stage is currently underway in our laboratories and it will be addressed in future works.

Further HPLC–DAD–MS/MS analysis of the extracts allowed the identification and quantification of three phenolic acids. The main compound detected was 4 hydroxybenzoic acid, with a content of 7.85 μg/g WIS, followed by salicylic and p-coumaric acids (0.51 and 0.12 μg/g WIS, respectively), as shown in Figure S1. This finding is in agreement with previous studies, where hydroxybenzoic acids were also predominant [8,58]. However, other compounds such as gallic acid, catechin or flavones, that were not found in autohydrolysis liquors, were reported in important quantities in melon by-product hydro-alcoholic extracts [8,14,66]. These differences could be associated with the melon variety, the state of ripening, the extraction techniques or analytical methodologies.

3.3. Pectin Recovery from Autohydrolysis Liquor

Pectin was recovered and purified from the autohydrolysis liquor obtained at 140 °C by precipitation using an ethanol concentration of 75% *v/v*. As shown in Table 2, the pectin yield was 16.59 g/100 g WIS, in dry weight, with a galacturonic acid content of 55.41% (91.23% of the galacturonic acid present in the original liquor). When comparing with the citric acid extraction, greatly differing melon peel pectin yields (3.24–29.48%) and lower galacturonic acid contents (indicating lower pectin purity) (47–48%) were reported [19,20]. Concerning similar raw materials, several autohydrolysis studies found a wide range in the galacturonic acid content of the obtained pectin, with higher values for the pomelo peels (76.62%) [54] and lower for apple pomace (48%) [55], both of which presented moderately higher pectin yields than this work (19.6 and 17.55%, respectively). Regarding other OSs, GalOS and AraOS also achieved good recoveries, both close to 70%, whereas the other oligosaccharides remained mostly in the liquid phase.

Table 2. Chemical characterization of the selected autohydrolysis liquor and the precipitated pectin.

	Autohydrolysis Liquor	Recovered Pectin	Recovery Yield
	Chemical Composition (g/100 g dw)		
NVC (g/100 g WIS)	23.22	16.59	71.43
GOS	2.50	0.14	4.07
XOS	1.40	0.00	0.00
GalOS	8.15	8.05	70.50
ManOS	1.97	0.00	0.00
AraOS	4.62	4.37	67.47
RhaOS	1.56	0.51	23.37
Ac-OS	2.05	1.57	54.68
OGalA	43.38	55.41	91.23
Protein	12.13	6.10	35.94
Ash	9.75	6.88	50.40
TPC	0.81	0.33	29.21
	Other Parameters		
Moisture (%)	98.65	7.02	
OS/MS ratio (*w/w*)	7.56	104.15	
DA (molar %)	15.26	9.15	
HG * (molar %)	62.34	77.75	
RG-I * (molar %)	27.71	22.03	
HG/RG-I	2.25	3.53	
	Sugar Ratios ** (*w/w*)		
1 $\frac{GalA}{Rha+Ara+Gal+Xyl}$	2.76	4.29	
2 $\frac{Rha}{GalA}$	0.04	0.01	
3 $\frac{Ara+Gal}{Rha}$	8.17	24.27	

* Determined as defined by M'sakni et al. [67]. ** Calculated according to Houben et al. [68]. GalA: galacturonic acid; Rha: rhamnose; Gal: galactose; Ara: arabinose; Xyl: xylose; HG: homogalacturonan; RG: rhamnogalacturonan.

The composition of the precipitated pectin can give more insight into the structure of the solubilized oligosaccharides. A recent study shows that hot water treatments are efficient for the extraction of the HG region of pectin with rhamnose contents of 0.5–0.6%, indicating a low RG-I content [69]. Our results confirm this finding, by increasing HGs and decreasing RG-I in the precipitated pectin, meaning that some of the branched oligosaccharides solubilized would be too degraded to be recovered by precipitation.

Moreover, the sugar ratios can also help to obtain information on the polymeric level [68,70], since sugar ratio 1 measures of the linearity of pectin, sugar ratio 2 indicates the contribution of RG-I to the pectin population and sugar ration 3 compares the amount of RG-I side-chain sugars to rhamnose [68]. In this case, the higher sugar ratio 1 in the precipitate proves the recuperation of the more linear region pectin, with a lower RG-I content (according to the sugar ratio 2 calculated value); whereas the high sugar ratio 3 indicates that the recovered RG region would be highly branched with GalOS and AraOS. Furthermore, the lack of XOS, ManOS and GOS shows that the co-extracted hemicelluloses and cellulosic oligomers were not precipitated with the pectin. A similar linearity and RG-I content, but less branching (sugar ratios of 5.9, 0.01 and 11.38) has also been reported in apple pomace pectin recovered by hydrothermal treatment at the same temperature [55].

Besides oligosaccharides, the recovered pectin contained a substantial amount of protein and some phenolic compounds (6.10 and 0.33 g/100 g dw, respectively), with recovery yields of 35.94 and 29.21%. Moreover, another important parameter to consider would be the OS/MS ratio, which increased from 7.56 to 104.15, with the recovered pectin having a monosaccharide concentration of just 0.67 g/100 g dw.

3.4. Structural Characterization

The properties of the recovered oligosaccharides are conditioned not only by their chemical composition but also by their structure, such as the degree of substitution or molecular weight distribution [34,71]. Therefore, a more detailed analysis of the final product can give more insight into its potential applications [71]. Several analytical techniques are often used in the literature to assess the effect of a treatment on the structure of the resulting oligosaccharides. For instance, FTIR and nuclear magnetic resonance (NMR) can be used to confirm the presence of characteristic functional groups or bonds [34,71,72], SEC allows for the determination of molecular weights and polydispersity index [26,32,70,71] and HPAEC–PAD and MALDI-TOF MS are commonly used to determine degrees of polymerization in oligosaccharides, with MALDI-TOF MS also being able to identify the composition of oligomers based on their molecular weight [26]. In this context, the liquors obtained in this study at 140 °C were freeze-dried and analyzed by complementary techniques: MALDI-TOF MS and HPAEC–PAD; and both the freeze-dried liquors and the pectin recovered by precipitation were analyzed by FTIR.

Both FTIR spectra (Figure 5) show similar bands to those of pectins found in the bibliography, especially the purified pectin with more intense peaks [19,20,31,70,72,73]. The peak at 3200–3500 cm^{-1} corresponds to OH stretching as a result of hydrogen bonding in pectin [19,31,71]. The peak between 2800 and 3000 cm^{-1} is related to stretching and bending vibrations of alkyl groups in the carbohydrates making up the pectin [20,31,71]. The observation of methyl esterified carboxyl groups (-$COOCH_3$, at 1738–1740 cm^{-1}) and free carboxyl groups (-COO^-, at 1605–1606 cm^{-1}) confirmed the presence of pectin with some degree of methyl esterification [19,31,72]. Using these two peaks and the calibration curve determined by Manrique and Lajolo [74], the degree of methyl esterification (DM) of the OGalA extracted in this work was estimated to be 41% (molar) for the autohydrolysis liquor and 50% for the recovered pectin. However, these values might actually be higher, since the considerable protein content of these extracts could be showing peaks at 1650 and 1539–1558 cm^{-1}, overlapping with the 1606 cm^{-1} band [70]. The increase in DM when recovering the pectin is in agreement with the literature regarding hot water treatments, which tend to extract pectins with high DM [69]. Free carboxyl groups also show weaker bands at 1410–1440 cm^{-1} related to symmetric stretching [19,73,75]. Other weaker and less relevant bands are the ones at 1370–1372 cm^{-1}, that could be caused by CH stretching vibrations [72,75] and the one at 1329–1332 cm^{-1}, that could be related to the COO- functional group in pectin [20]. The patterns between 1200 and 800 cm^{-1} represent the "finger print" area, which is unique to a compound, making its interpretation a complicated matter [19,73].

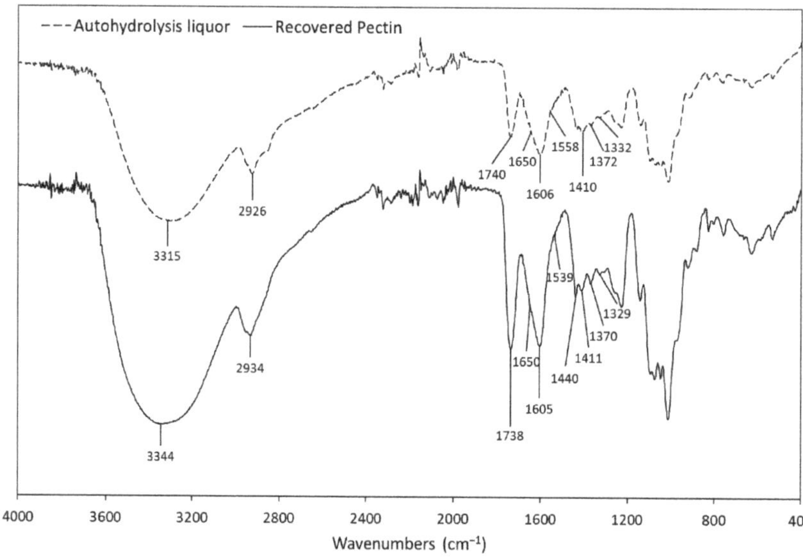

Figure 5. FTIR spectra recorded for the autohydrolysis liquors and recovered pectin.

The degree of polymerization (DP) of the oligogalacturonides was determined by MALDI-TOF MS. Table 3 displays the mass signals in MALDI-TOF spectra and suggested oligomeric structures (all compounds were identified as sodium or potassium adducts). The MALDI-TOF MS data confirmed the presence of a wide variety of complex oligomers, with the main products being combinations of galaturonic acid in the range of 1–11 with neutral sugars (mainly hexoses, pentoses and rhamnose) and substituted by acetyl and methyl groups. Overall, the complex chemical structure determined for the oligomeric compounds obtained from the melon peels is similar to what it has been reported for other oligosaccharides obtained by enzymatic hydrolysis from lemon peels [76]. Furthermore, HPAEC–PAD analysis (Figure 6) showed OGalA with DP from 5 to over 20 (using commercial polygalacturonic acid (PGA) as standard).

Table 3. MALDI-TOF results and suggested structures of autohydrolysis liquors at 140 °C.

m/z	Structure
648.52	$Pent_2GalAAc_4Na$
847.58	$HexPentRhaGalA_2MeNa$
980.57	$Pent_3GalA_3MeNa$
1174.34	$HexGalA_4Me_4Ac_5Na$
1463.65	$Hex_2GalA_6Me_3Na$
1603.34	$GalA_8Me_5Ac_2Na$
1790.94	$Pent_2GalA_7Me_3Ac_5Na$
1889.26	$Pent_3GalA_8Me_3Na$
2409.23	$Hex_1Pent_3GalA_9Me_4Ac_4Na$
2641.94	$Hex_4PentGalA_{10}AcK$
3498.19	$Hex_4Pent_2Rha_3GalA_{11}Me_3Ac_3Na$

Hex: hexose; Pent: pentose; Rha: rhamnose; Ac: acetyl group; GalA: galacturonic acid; Me: methyl group.

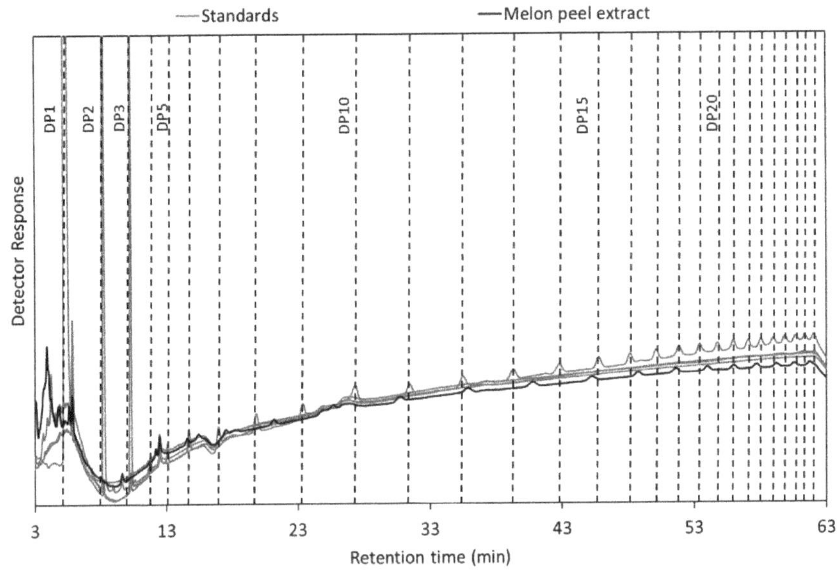

Figure 6. HPAEC–PAD of melon peel extracts and standards (DP1, DP2, DP3 and PGA).

3.5. Valorization of the Autohydrolysis Spent Solids

Melon peels can be considered an abundant, renewable and cheap source of biomass whose complete valorization to obtain various biobased products is an interesting strategy from an economic and environmental point of view. In this context, the potential application of the spent solid recovered from the selected hydrothermal treatment has been evaluated. It should be noted that operating at 140 °C, the solid yield was 72.88%, hence the importance of its valorization. The main components of the autohydrolyzed solid were glucan and Klason lignin, with values of 31.26 and 30.59 g/100 g spent solid, respectively, and percentages of recovery for both fractions of 93% and close to 100%. These results are better than those previously reported for the autohydrolysis of other raw materials rich in pectin. When sugar beet pulp and orange and lemon peels were treated under similar severities, the percentages of glucan recoveries were in the range of 75–80% and in the case of citric peels, lignin recoveries close to 92% were achieved [27–29]. The selected operational temperature allowed reaching high pectic oligosaccharide extraction, whereas, as could be expected, other hemicellulosic fractions remained in the spent solid; since according to the literature their solubilization requires higher temperatures [34,52,77]. This leads to the fact that small amounts of xylan, galacturonan and galactan were also detected in the autohydrolyzed solid (6.90, 5.79 and 2.78%, respectively), which are followed by acetyl groups, mannan and arabinosyl substituents, with values lower than 2%. The behavior pattern of these compounds during autohydrolysis could be related with structural features, since they could be part of some hemicellulosic polysaccharides or be linked to the galacturonic acid of pectin.

Taking into account the chemical composition of the spent solid recovered in the autohydrolysis of melon peels at 140 °C, it would be plausible to implement a second autohydrolysis step to obtain a solid fraction enriched in cellulose with potential applications, and a liquid fraction with a higher phenolic content [78]. For instance, this cellulose would be suitable for the biotechnological production of fermentable sugars, or for the production of nanocellulose [13]. In this work, the digestibility of the spent solid was assessed without further processing, by subjecting it to an enzymatic hydrolysis using an enzyme cocktail, specifically a mixture of endopolygalacturonase, cellulase and β-glucosidase. Under the experimental conditions selected (see in the Materials and Methods Section), after 24 h of saccharification, 79.83% of the glucan was solubilized, yielding a solution with 22.41 g/L of glucose.

Other compounds, such as galactose and galacturonic acid, were also detected in the reaction medium, but in much lower amounts (3.29 and 2.7 g/L, respectively). Further increases in the hydrolysis time led to glucan conversions into glucose up to 87.65% after 48 h.

4. Conclusions

Pectic oligosaccharides with antioxidant activity were recovered from melon by-products using an environmentally friendly process. The raw material was subjected to a water extraction to release free sugars and the remaining solid was treated by autohydrolysis in non-isothermal conditions. The selected temperature for this treatment was 140 °C (severity of 2.03), achieving a liquor with a total oligosaccharide content of 15.24 g/100 g WIS, of which 10.07 g/100 g WIS were OGalA. A structural characterization confirmed the presence of OGalA with a degree of methyl esterification of at least 41% and a wide range of polymerization degrees. These liquors also exhibited antioxidant activity and contained a considerable amount of protein (2.82 g/100 g WIS), with a negligible amount of non-volatile impurities. The pectic oligosaccharides were successfully recovered by precipitation, producing a mostly linear pectin with 55.41% of galacturonic acid. In conclusion, an environmentally friendly process was developed for the valorization of melon by-products, through the production of pectic oligosaccharides with protein content and antioxidant activity, and potential applications in the food and pharmaceutical industries.

Supplementary Materials: The following are available online at http://www.mdpi.com/2304-8158/9/11/1702/s1, Figure S1: HPLC UV/VIS chromatograms of melon peel autohydrolysis liquors at 140 °C.

Author Contributions: Conceptualization, X.R., B.G. and R.Y.; methodology, X.R., B.G. and R.Y.; software, X.R.; validation, B.G. and R.Y.; formal analysis, X.R., B.G. and R.Y.; investigation, X.R.; resources, B.G. and R.Y.; data curation, X.R.; writing—original draft preparation, X.R.; writing—review and editing, B.G. and R.Y.; visualization, X.R., B.G. and R.Y.; supervision, B.G. and R.Y.; project administration, B.G. and R.Y.; funding acquisition, B.G. and R.Y. All authors have read and agreed to the published version of the manuscript.

Funding: The authors acknowledge the financial support received from the Spanish "Ministry of Economy and Competitiveness" (Project "Modified aqueous media for wood biorefineries", reference CTQ2017- 82962-R) and from "Xunta de Galicia" (GRC ED431C 2018/47, Centro Singular de Investigación Biomédica "CINBIO" and Project ED431F 2020/03). These projects are partially funded by the FEDER Program of the European Union ("Unha maneira de facer Europa"). Beatriz Gullón would like to express her gratitude to the Spanish Ministry of Economy and Competitiveness for her postdoctoral grant (Reference RYC2018-026177-I). Xiana Rico is grateful for her predoctoral grant (reference ED481A-2018/300) obtained with financial support from the "Xunta de Galicia".

Conflicts of Interest: The authors declare no conflict of interest. The funders had no role in the design of the study; in the collection, analyses, or interpretation of data; in the writing of the manuscript, or in the decision to publish the results.

References

1. FAO. The State of Food and Agriculture 2019. In *Moving Forward on Food Loss and Waste Reduction*; FAO: Rome, Italy, 2019.
2. Barilla Center for Food & Nutrition. *Food Waste: Causes, Impacts and Proposals*; Barilla Center for Food & Nutrition: Parma, Italy, 2012.
3. Stenmarck, Å.; Jensen, C.; Quested, T.; Moates, G. *Estimates of European Food Waste Levels*; IVL Swedish Environmental Research Institute: Stockholm, Sweden, 2016.
4. Fava, F.; Totaro, G.; Diels, L.; Reis, M.; Duarte, J.; Poggi-varaldo, M.; Ferreira, B.S.; Carioca, O.B. Biowaste biorefinery in Europe: Opportunities and research & development needs. *New Biotechnol.* **2015**, *32*, 100–108. [CrossRef]
5. Banerjee, J.; Singh, R.; Vijayaraghavan, R.; Macfarlane, D.; Patti, A.F.; Arora, A. Bioactives from fruit processing wastes: Green approaches to valuable chemicals. *Food Chem.* **2017**, *225*, 10–22. [CrossRef] [PubMed]
6. European Commission. *Directive (EU) 2018/850 Amending Directive 1999/31/EC on the Landfill of Waste*; OJ L 150; European Commission: Brussels, Belgium, 2018; pp. 100–108.

7. European Commission. *Directive (EU) 2018/851 Amending Directive 2008/98/EC on Waste*; OJ L 150; European Commission: Brussels, Belgium, 2018; pp. 109–140.
8. Mallek-Ayadi, S.; Bahloul, N.; Kechaou, N. Characterization, phenolic compounds and functional properties of *Cucumis melo* L. peels. *Food Chem.* **2017**, *221*, 1691–1697. [CrossRef] [PubMed]
9. FAO. Faostat Database. Available online: www.fao.org/faostat (accessed on 1 April 2020).
10. Aguayo, E.; Escalona, V.H.; Artés, F.A. Metabolic Behavior and Quality Changes of Whole and Fresh Processed Melon. *J. Food Sci.* **2004**, *69*, 148–155. [CrossRef]
11. Fundo, J.F.; Miller, F.A.; Garcia, E.; Santos, J.R.; Silva, C.L.M.; Brandão, T.R.S. Physicochemical characteristics, bioactive compounds and antioxidant activity in juice, pulp, peel and seeds of Cantaloupe melon. *J. Food Meas. Charact.* **2018**, *12*, 292–300. [CrossRef]
12. Gómez-García, R.; Campos, D.A.; Aguilar, C.N.; Madureira, A.R.; Pintado, M. Valorization of melon fruit (*Cucumis melo* L.) by-products: Phytochemical and Biofunctional properties with Emphasis on Recent Trends and Advances. *Trends Food Sci. Technol.* **2020**, *99*, 507–519. [CrossRef]
13. Rico, X.; Gullón, B.; Alonso, J.L.; Yáñez, R. Recovery of high value-added compounds from pineapple, melon, watermelon and pumpkin processing by-products: An overview. *Food Res. Int.* **2020**, *132*, 109086. [CrossRef]
14. Rolim, P.M.; Fidelis, G.P.; Padilha, C.E.A.; Santos, E.S.; Rocha, H.A.O.; Macedo, G.R. Phenolic profile and antioxidant activity from peels and seeds of melon (*Cucumis melo* L. var. *reticulatus*) and their antiproliferative effect in cancer cells. *Braz. J. Med. Biol. Res.* **2018**, *51*, 1–14. [CrossRef]
15. Rolim, P.M.; de Oliveira Júnior, S.D.; Mendes de Oliveira, A.C.S.; dos Santos, E.S.; de Macedo, G.R. Nutritional value, cellulase activity and prebiotic effect of melon residues (*Cucumis melo* L. *reticulatus* group) as a fermentative substrate. *J. Food Nutr. Res.* **2018**, *57*, 315–327.
16. Toledo, N.M.V.; Mondoni, J.; Harada-Padermo, S.S.; Vela-Paredes, R.S.; Berni, P.R.A.; Selani, M.M.; Canniatti Brazaca, S.G. Characterization of apple, pineapple, and melon by-products and their application in cookie formulations as an alternative to enhance the antioxidant capacity. *J. Food Process. Preserv.* **2019**, *43*, e14100. [CrossRef]
17. Wang, F.; Li, H.; Zhao, H.; Zhang, Y.; Qiu, P.; Li, J.; Wang, S. Antidiabetic Activity and Chemical Composition of Sanbai Melon Seed Oil. Evidence-Based Complement. *Altern. Med.* **2018**, *2018*, 1–14. [CrossRef]
18. Silva, M.A.; Albuquerque, T.G.; Alves, R.C.; Oliveira, M.B.P.P.; Costa, H.S. Melon (*Cucumis melo* L.) by-products: Potential food ingredients for novel functional foods? *Trends Food Sci. Technol.* **2020**, *98*, 181–189. [CrossRef]
19. Raji, Z.; Khodaiyan, F.; Rezaei, K.; Kiani, H.; Hosseini, S.S. Extraction optimization and physicochemical properties of pectin from melon peel. *Int. J. Biol. Macromol.* **2017**, *98*, 709–716. [CrossRef] [PubMed]
20. Muthukumaran, C.; Banupriya, L.; Harinee, S.; Sivaranjani, S.; Sharmila, G.; Rajasekar, V.; Kumar, N.M. Pectin from muskmelon (*Cucumis melo* var. *reticulatus*) peels: Extraction optimization and physicochemical properties. *3 Biotech.* **2017**, *7*, 66. [CrossRef]
21. Naqash, F.; Masoodi, F.A.; Rather, S.A.; Wani, S.M.; Gani, A. Emerging concepts in the nutraceutical and functional properties of pectin—A Review. *Carbohydr. Polym.* **2017**, *168*, 227–239. [CrossRef]
22. Adetunji, L.R.; Adekunle, A.; Orsat, V.; Raghavan, V. Advances in the pectin production process using novel extraction techniques: A review. *Food Hydrocoll.* **2017**, *62*, 239–250. [CrossRef]
23. Gullón, B.; Gómez, B.; Martínez-Sabajanes, M.; Yáñez, R.; Parajó, J.C.; Alonso, J.L. Pectic oligosaccharides: Manufacture and functional properties. *Trends Food Sci. Technol.* **2013**, *30*, 153–161. [CrossRef]
24. Wang, W.; Chen, W.; Zou, M.; Lv, R.; Wang, D.; Hou, F.; Feng, H.; Ma, X.; Zhong, J.; Ding, T.; et al. Applications of power ultrasound in oriented modification and degradation of pectin: A review. *J. Food Eng.* **2018**, *234*, 98–107. [CrossRef]
25. Kumar, M.; Tomar, M.; Saurabh, V.; Mahajan, T.; Punia, S.; Contreras, M.; del Mar Contreras, M.; Rudra, S.G.; Kaur, C.; Kennedy, J.F. Emerging trends in pectin extraction and its anti-microbial functionalization using natural bioactives for application in food packaging. *Trends Food Sci. Technol.* **2020**, *105*, 223–237. [CrossRef]
26. Babbar, N.; Dejonghe, W.; Gatti, M.; Sforza, S.; Elst, K. Pectic oligosaccharides from agricultural by-products: Production, characterization and health benefits. *Crit. Rev. Biotechnol.* **2015**, *36*, 1–13. [CrossRef]
27. Gómez, B.; Gullón, B.; Yáñez, R.; Parajó, J.C.; Alonso, J.L. Pectic Oligosacharides from Lemon Peel Wastes: Production, Purification, and Chemical Characterization. *J. Agric. Food Chem.* **2013**, *61*, 10043–10053. [CrossRef] [PubMed]

28. Martínez, M.; Yáñez, R.; Alonso, J.L.; Parajó, J.C. Chemical Production of Pectic Oligosaccharides from Orange Peel Wastes. *Ind. Eng. Chem. Res.* **2010**, *49*, 8470–8476. [CrossRef]
29. Martínez, M.; Gullón, B.; Schols, H.A.; Alonso, J.L.; Parajó, J.C. Assessment of the Production of Oligomeric Compounds from Sugar Beet Pulp. *Ind. Eng. Chem. Res.* **2009**, *48*, 4681–4687. [CrossRef]
30. Talekar, S.; Patti, A.F.; Vijayaraghavan, R.; Arora, A. An integrated green biorefinery approach towards simultaneous recovery of pectin and polyphenols coupled with bioethanol production from waste pomegranate peels. *Bioresour. Technol.* **2018**, *266*, 322–334. [CrossRef]
31. Banerjee, J.; Singh, R.; Vijayaraghavan, R.; MacFarlane, D.; Patti, A.F.; Arora, A. A hydrocolloid based biorefinery approach to the valorisation of mango peel waste. *Food Hydrocoll.* **2018**, *77*, 142–151. [CrossRef]
32. Dávila, I.; Gordobil, O.; Labidi, J.; Gullón, P. Assessment of suitability of vine shoots for hemicellulosic oligosaccharides production through aqueous processing. *Bioresour. Technol.* **2016**, *211*, 636–644. [CrossRef]
33. Garrote, G.; Domínguez, H.; Parajó, J.C. Mild autohydrolysis: An environmentally friendly technology for xylooligosaccharide production from wood. *J. Chem. Technol. Biotechnol.* **1999**, *74*, 1101–1109. [CrossRef]
34. Dávila, I.; Gullón, B.; Alonso, J.L.; Labidi, J.; Gullón, P. Vine shoots as new source for the manufacture of prebiotic oligosaccharides. *Carbohydr. Polym.* **2019**, *207*, 34–43. [CrossRef]
35. López, M.; Penín, L.; Vila, C.; Santos, V.; Parajó, J.C. Multi-Stage Hydrothermal Processing of Eucalyptus Globulus Wood: An Experimental Assessment. *J. Wood Chem. Technol.* **2019**, *39*, 329–342. [CrossRef]
36. Grajek, W.; Olejnik, A.; Sip, A. Probiotics, prebiotics and antioxidants as functional foods. *Acta Biochim. Pol.* **2005**, *52*, 665–671. [CrossRef]
37. Willats, W.G.T.; Knox, J.P.; Mikkelsen, J.D. Pectin: New insights into an old polymer are starting to gel. *Trends Food Sci. Technol.* **2006**, *17*, 97–104. [CrossRef]
38. De Paulo Farias, D.; de Araújo, F.F.; Neri-Numa, I.A.; Pastore, G.M. Prebiotics: Trends in food, health and technological applications. *Trends Food Sci. Technol.* **2019**, *93*, 23–35. [CrossRef]
39. Sagar, N.A.; Pareek, S.; Sharma, S.; Yahia, E.M.; Lobo, M.G. Fruit and Vegetable Waste: Bioactive Compounds, Their Extraction, and Possible Utilization. *Compr. Rev. Food Sci. Food Saf.* **2018**, *17*, 512–531. [CrossRef]
40. Morais, D.R.; Rotta, E.M.; Sargi, S.C.; Schmidt, E.M.; Bonafe, E.G.; Eberlin, M.N.; Sawaya, A.C.H.F.; Visentainer, J.V. Antioxidant activity, phenolics and UPLC-ESI(-)-MS of extracts from different tropical fruits parts and processed peels. *Food Res. Int.* **2015**, *77*, 392–399. [CrossRef]
41. Guo, C.; Yang, J.; Wei, J.; Li, Y.; Xu, J.; Jiang, Y. Antioxidant activities of peel, pulp and seed fractions of common fruits as determined by FRAP assay. *Nutr. Res.* **2003**, *23*, 1719–1726. [CrossRef]
42. Ismail, H.I.; Chan, K.W.; Mariod, A.A.; Ismail, M. Phenolic content and antioxidant activity of cantaloupe (cucumis melo) methanolic extracts. *Food Chem.* **2010**, *119*, 643–647. [CrossRef]
43. Miller, F.A.; Fundo, J.F.; Garcia, E.; Santos, J.R.; Silva, C.L.M.; Brandão, T.R.S. Physicochemical and Bioactive Caracterisation of Edible and Waste Parts of "Piel de Sapo" Melon. *Horticulturae* **2020**, *6*, 60. [CrossRef]
44. Míguez, B.; Gómez, B.; Gullón, P.; Gullón, B.; Alonso, J.L. Pectic Oligosaccharides and Other Emerging Prebiotics. In *Probiotics and Prebiotics in Human Nutrition and Health*; InTech: London, UK, 2016; pp. 267–322.
45. Lavoie, J.M.; Capek-Menard, E.; Gauvin, H.; Chornet, E. Quality pulp from mixed softwoods as an added value coproduct of a biorefinery. *Ind. Eng. Chem. Res.* **2010**, *49*, 2503–2509. [CrossRef]
46. Blumenkrantz, N.; Asboe-Hansen, G. New method for quantitative determination of uronic acids. *Anal. Biochem.* **1973**, *54*, 484–489. [CrossRef]
47. Martínez, M.; Gullón, B.; Yáñez, R.; Alonso, J.L.; Parajó, J.C. Direct Enzymatic Production of Oligosaccharide Mixtures from Sugar Beet Pulp: Experimental Evaluation and Mathematical Modeling. *J. Agric. Food Chem.* **2009**, *57*, 5510–5517. [CrossRef]
48. Singleton, V.L.; Rossi, J.A. Colorimetry of Total Phenolics with Phosphomolybdic-Phosphotungstic Acid Reagents. *Am. J. Enol. Vitic.* **1965**, *16*, 144–158.
49. Gullón, B.; Eibes, G.; Moreira, M.T.; Dávila, I.; Labidi, J.; Gullón, P. Antioxidant and antimicrobial activities of extracts obtained from the refining of autohydrolysis liquors of vine shoots. *Ind. Crops Prod.* **2017**, *107*, 105–113. [CrossRef]
50. Gómez, B.; Gullón, B.; Remoroza, C.; Schols, H.A.; Parajó, J.C.; Alonso, J.L. Purification, characterization, and prebiotic properties of pectic oligosaccharides from orange peel wastes. *J. Agric. Food Chem.* **2014**, *62*, 9769–9782. [CrossRef] [PubMed]

51. Morais, D.R.; Rotta, E.M.; Sargi, S.C.; Bonafe, E.G.; Suzuki, R.M.; Souza, N.E.; Matsushita, M.; Visentainer, J.V. Proximate composition, mineral contents and fatty acid composition of the different parts and dried peels of tropical fruits cultivated in Brazil. *J. Braz. Chem. Soc.* **2017**, *28*, 308–318. [CrossRef]
52. Rico, X.; Gullón, B.; Alonso, J.L.; Parajó, J.C.; Yáñez, R. Valorization of peanut shells: Manufacture of bioactive oligosaccharides. *Carbohydr. Polym.* **2018**, *183*, 21–28. [CrossRef]
53. Klinchongkon, K.; Khuwijitjaru, P.; Wiboonsirikul, J.; Adachi, S. Extraction of Oligosaccharides from Passion Fruit Peel by Subcritical Water Treatment. *J. Food Process Eng.* **2015**, *40*, e12269. [CrossRef]
54. Liew, S.Q.; Teoh, W.H.; Tan, C.K.; Yusoff, R.; Ngoh, G.C. Subcritical water extraction of low methoxyl pectin from pomelo (*Citrus grandis* (L.) Osbeck) peels. *Int. J. Biol. Macromol.* **2018**, *116*, 128–135. [CrossRef]
55. Wang, X.; Lü, X. Characterization of pectic polysaccharides extracted from apple pomace by hot-compressed water. *Carbohydr. Polym.* **2014**, *102*, 174–184. [CrossRef]
56. Muñoz-Almagro, N.; Valadez-Carmona, L.; Mendiola, J.A.; Ibáñez, E.; Villamiel, M. Structural characterisation of pectin obtained from cacao pod husk. Comparison of conventional and subcritical water extraction. *Carbohydr. Polym.* **2019**, *217*, 69–78. [CrossRef]
57. Li, W.J.; Fan, Z.G.; Wu, Y.Y.; Jiang, Z.G.; Shi, R.C. Eco-friendly extraction and physicochemical properties of pectin from jackfruit peel waste with subcritical water. *J. Sci. Food Agric.* **2019**, *99*, 5283–5292. [CrossRef]
58. Pérez-Jiménez, J.; Saura-Calixto, F. Fruit peels as sources of non-extractable polyphenols or macromolecular antioxidants: Analysis and nutritional implications. *Food Res. Int.* **2018**, *111*, 148–152. [CrossRef] [PubMed]
59. Pérez-Jiménez, J.; Saura-Calixto, F. Macromolecular antioxidants or non-extractable polyphenols in fruit and vegetables: Intake in four European countries. *Food Res. Int.* **2015**, *74*, 315–323. [CrossRef] [PubMed]
60. Ballesteros, L.F.; Ramirez, M.J.; Orrego, C.E.; Teixeira, J.A.; Mussatto, S.I. Optimization of autohydrolysis conditions to extract antioxidant phenolic compounds from spent coffee grounds. *J. Food Eng.* **2017**, *199*, 1–8. [CrossRef]
61. Pérez-Armada, L.; Rivas, S.; González, B.; Moure, A. Extraction of phenolic compounds from hazelnut shells by green processes. *J. Food Eng.* **2019**, *255*, 1–8. [CrossRef]
62. Gullón, P.; Eibes, G.; Lorenzo, J.M.; Pérez-Rodríguez, N.; Lú-Chau, T.A.; Gullón, B. Green sustainable process to revalorize purple corn cobs within a biorefinery frame: Co-production of bioactive extracts. *Sci. Total Environ.* **2020**, *709*, 136236. [CrossRef] [PubMed]
63. Leschinsky, M.; Zuckerstätter, G.; Weber, H.K.; Patt, R.; Sixta, H. Effect of autohydrolysis of Eucalyptus globulus wood on lignin structure. Part 1: Comparison of different lignin fractions formed during water prehydrolysis. *Holzforschung* **2008**, *62*, 645–652. [CrossRef]
64. Ballesteros, L.F.; Teixeira, J.A.; Mussatto, S.I. Extraction of polysaccharides by autohydrolysis of spent coffee grounds and evaluation of their antioxidant activity. *Carbohydr. Polym.* **2017**, *157*, 258–266. [CrossRef]
65. Gullón, B.; Eibes, G.; Dávila, I.; Moreira, M.T.; Labidi, J.; Gullón, P. Hydrothermal treatment of chestnut shells (*Castanea sativa*) to produce oligosaccharides and antioxidant compounds. *Carbohydr. Polym.* **2018**, *192*, 75–83. [CrossRef]
66. Vella, F.M.; Cautela, D.; Laratta, B. Characterization of Polyphenolic Compounds in Cantaloupe Melon By-Products. *Foods* **2019**, *8*, 196. [CrossRef]
67. M'sakni, N.H.; Majdoub, H.; Roudesli, S.; Picton, L.; Le Cerf, D.; Rihouey, C.; Morvan, C. Composition, structure and solution properties of polysaccharides extracted from leaves of Mesembryanthenum crystallinum. *Eur. Polym. J.* **2006**, *42*, 786–795. [CrossRef]
68. Houben, K.; Jolie, R.P.; Fraeye, I.; Van Loey, A.M.; Hendrickx, M.E. Comparative study of the cell wall composition of broccoli, carrot, and tomato: Structural characterization of the extractable pectins and hemicelluloses. *Carbohydr. Res.* **2011**, *346*, 1105–1111. [CrossRef] [PubMed]
69. Mao, G.; Wu, D.; Wei, C.; Tao, W.; Ye, X.; Linhardt, R.J.; Orfila, C.; Chen, S. Reconsidering conventional and innovative methods for pectin extraction from fruit and vegetable waste: Targeting rhamnogalacturonan I. *Trends Food Sci. Technol.* **2019**, *94*, 65–78. [CrossRef]
70. Kpodo, F.M.; Agbenorhevi, J.K.; Alba, K.; Bingham, R.J.; Oduro, I.N.; Morris, G.A.; Kontogiorgos, V. Pectin isolation and characterization from six okra genotypes. *Food Hydrocoll.* **2017**, *72*, 323–330. [CrossRef]
71. Grassino, A.N.; Barba, F.J.; Brnčić, M.; Lorenzo, J.M.; Lucini, L.; Brnčić, S.R. Analytical tools used for the identification and quantification of pectin extracted from plant food matrices, wastes and by-products: A review. *Food Chem.* **2018**, *266*, 47–55. [CrossRef]

72. Kazemi, M.; Khodaiyan, F.; Labbafi, M.; Saeid Hosseini, S.; Hojjati, M. Pistachio green hull pectin: Optimization of microwave-assisted extraction and evaluation of its physicochemical, structural and functional properties. *Food Chem.* **2019**, *271*, 663–672. [CrossRef] [PubMed]
73. Hosseini, S.S.; Khodaiyan, F.; Yarmand, M.S. Optimization of microwave assisted extraction of pectin from sour orange peel and its physicochemical properties. *Carbohydr. Polym.* **2016**, *140*, 59–65. [CrossRef] [PubMed]
74. Manrique, G.D.; Lajolo, F.M. FT-IR spectroscopy as a tool for measuring degree of methyl esterification in pectins isolated from ripening papaya fruit. *Postharvest Biol. Technol.* **2002**, *25*, 99–107. [CrossRef]
75. Gnanasambandam, R.; Proctor, A. Determination of pectin degree of esterification by diffuse reflectance Fourier transform infrared spectroscopy. *Food Chem.* **2000**, *68*, 327–332. [CrossRef]
76. Gómez, B.; Yáñez, R.; Parajó, J.C.; Alonso, J.L. Production of pectin-derived oligosaccharides from lemon peels by extraction, enzymatic hydrolysis and membrane filtration. *J. Chem. Technol. Biotechnol.* **2016**, *91*, 234–247. [CrossRef]
77. Penín, L.; Santos, V.; del Río, J.C.; Parajó, J.C. Assesment on the chemical fractionation of Eucalyptus nitens wood: Characterization of the products derived from the structural components. *Bioresour. Technol.* **2019**, *281*, 269–276. [CrossRef]
78. Bassani, A.; Fiorentini, C.; Vadivel, V.; Moncalvo, A.; Spigno, G. Implementation of auto-hydrolysis process for the recovery of antioxidants and cellulose from wheat straw. *Appl. Sci.* **2020**, *10*, 6112. [CrossRef]

foods

MDPI

Review

By-Products of Agri-Food Industry as Tannin-Rich Sources: A Review of Tannins' Biological Activities and Their Potential for Valorization

María Fraga-Corral [1,2], Paz Otero [1,3], Javier Echave [1], Paula Garcia-Oliveira [1,2], Maria Carpena [1], Amira Jarboui [1], Bernabé Nuñez-Estevez [1,2], Jesus Simal-Gandara [1,*] and Miguel A. Prieto [1,2,*]

1 Nutrition and Bromatology Group, Analytical and Food Chemistry Department, Faculty of Food Science and Technology, Ourense Campus, University of Vigo, 32004 Ourense, Spain; mfraga@uvigo.es (M.F.-C.); pazoterofuertes@gmail.com (P.O.); ecalja@outlook.es (J.E.); paula.garcia.oliveira@uvigo.es (P.G.-O.); maria.carpena.rodriguez@uvigo.es (M.C.); jarb.amira@gmail.com (A.J.); bernabenunez16@gmail.com (B.N.-E.)
2 Centro de Investigação de Montanha (CIMO), Campus de Santa Apolonia, Instituto Politécnico de Bragança, 5300-253 Bragança, Portugal
3 Department of Pharmacology, Pharmacy and Pharmaceutical Technology, Faculty of Veterinary, University of Santiago of Compostela, 27002 Lugo, Spain
* Correspondence: jsimal@uvigo.es (J.S.-G.); mprieto@uvigo.es (M.A.P.)

Abstract: During recent decades, consumers have been continuously moving towards the substitution of synthetic ingredients of the food industry by natural products, obtained from vegetal, animal or microbial sources. Additionally, a circular economy has been proposed as the most efficient production system since it allows for reducing and reutilizing different wastes. Current agriculture is responsible for producing high quantities of organic agricultural waste (e.g., discarded fruits and vegetables, peels, leaves, seeds or forestall residues), that usually ends up underutilized and accumulated, causing environmental problems. Interestingly, these agri-food by-products are potential sources of valuable bioactive molecules such as tannins. Tannins are phenolic compounds, secondary metabolites of plants widespread in terrestrial and aquatic natural environments. As they can be found in plenty of plants and herbs, they have been traditionally used for medicinal and other purposes, such as the leather industry. This fact is explained by the fact that they exert plenty of different biological activities and, thus, they entail a great potential to be used in the food, nutraceutical and pharmaceutical industry. Consequently, this review article is directed towards the description of the biological activities exerted by tannins as they could be further extracted from by-products of the agri-food industry to produce high-added-value products.

Keywords: tannins; valorization; circular economy; biological properties; health benefits

1. Introduction

1.1. Tannins as Target Compounds

Tannins are a diverse group within phenolic compounds widely distributed in nature. They are secondary metabolites of plants usually produced as a result of stress and they exert a protective role, including photoprotection against UV rays and free radicals or defense against other organisms and environmental conditions, such as dryness [1–3]. Tannins are a heterogeneous group, having molecular weights between 500 and 20,000 Da and very different chemical structures [4]. Tannins have been demonstrated to exert different biological activities, such as antioxidant activity. This property is related to their chemical structure as they possess phenolic rings able to bind to a wide range of molecules and act as electron scavengers to trap ions and radicals [2,4]. Generally, tannins possess about 12–16 phenolic groups and five to seven aromatic rings per 1000 Da [5]. They also present plenty of hydroxyl groups, which confer on them hydrophilic properties, solubility in aqueous solvents and also the ability to form complexes with proteins, carbohydrates, nucleic acids and alkaloids [6,7]. Regarding tannin classification, they have been historically classified into hydrolyzable tannins (HTs) and condensed tannins (CTs), and the latter are also called proanthocyanidins. Nowadays, the classification according to their chemical characteristic and structural properties has been updated. Thus, tannins can be grouped into gallotannins, ellagitannins, CTs, complex tannins (CoTs) and phlorotannins (PTs, an exclusive class of tannins found in the algal species of the Phaeophyceae class) [1,2,8]. A schematic representation of tannin structural classification is presented in Figure 1.

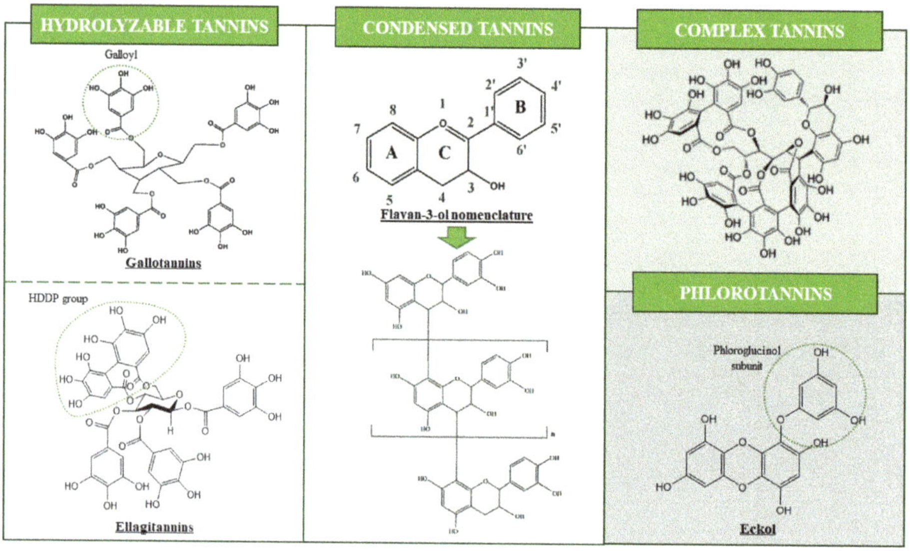

Figure 1. Structural classification of tannins. Functional groups are shown in circles.

HTs have owned their name as they can be hydrolyzed by weak acids/bases, producing carbohydrates and phenolic acids because of the reaction [3]. They are formed by glycosylated gallic acid units [9], which can be either ellagic acid (EA) or gallic acid (GA), forming ellagitannins (ETs) and gallotannins (GTs), respectively [3,10]. ETs are formed by simple to multiple units of hexahydroxydiphenol (HHDP) connected to a polyol core. After hydrolysis and the breakdown of C-C bonds between suitably orientated galloyl residues of glucogalloyl molecules of HHDP, they are converted into EA units [2,11,12]. The abundant variety of structure has been observed within this group, due to the different possibilities in the formation of oxidative linkages [7]. On the other hand, GTs are considered as simpler HTs and are formed by galloyl or digalloyl units coupled to a polyol, catechin or triterpenoid unit, in the form of pentagalloyl glucose (PGG). They can yield GA from the hydrolysis reaction [2,13]. HTs can be mainly found in fruits, berries, legumes, leafy vegetables and different tree species [3,9]. They have been widely employed in the leather industry and they have been studied for their antioxidant and antimicrobial properties [3].

CTs or proanthocyanidins account for more than 90% of the world commercial production of tannins [3]. They are polymeric or oligomeric flavan-3-ols, formed by the combination of A (phloroglucinol or resorcinol) and B (catechol or pyrogallol) rings [2,14] (Figure 1). When these compounds are heated in ethanol solutions in acidic conditions, they are decomposed into anthocyanidins [7]. The combination of these flavan-3-ol monomers gives rise to the formation of procyanidins (PCs) (composed of catequins and epicatechins) or profisetinidin, prorobinetidin and prodelphindin (composed of (epi)fisetinidol, (epi)robinetinidol and (epi)gallocatechin units) [2,15]. Particularly, these type of tannins are commonly found in fruits, berries, cocoa and some drinks such as wine, beer or tea [9].

Finally, CoTs and PTs will be briefly described. CoTs are tannins of high molecular weight, created as a result of the bonding between flavan-3-ols with either GTs or ETs. This type of tannin can be obtained from tree species such as *Quercus* sp. and *Castanea sativa* [6]. Finally, PTs are polyphenols obtained from brown marine algae and formed by phloroglucinol (1,3,5-trihydroxybenzene) (PG) synthesized via the acetate–malonate pathway. They are grouped into six major groups (fucols, phloroetols, fucophloroteols, fuhalols, carmalols and eckols) depending on to the type of bonds between PG units and their content of hydroxyl groups, being more complex with a higher level of PG units [16,17]. These tannins, which can represent up to 30% of seaweeds' dry weight, have been demonstrated to exert antimicrobial, photoprotection or antioxidant activities, among others [17,18].

Although tannins have been sometimes linked to unpleasant organoleptic properties, they have also shown plenty of properties and applications. Some of these properties are antioxidant, antimicrobial or anti-inflammatory, among others, which have given rise to their use in the food, nutraceutical and pharmaceutical industry [19]. Additionally, their toxic effects have been assessed [1]. Particularly, they have been proposed as natural food additives able to enhance the safety and the shelf life of products and also as clarification agents in drinks [20]. Furthermore, tannins have been used as adhesives and coatings, foams or adsorbents, among many other applications [3]. Different species containing diverse tannins have been included as part of patents that aim to exploit their properties and to create innovative applications (Table 1).

Table 1. Examples of patented tannin applications.

Tannins	Properties	Patent No.
Punicalin, punicalagin, pedunculagin, tellimagrandin, corilagin, granatine a and b, terminalin	Treatment or prevention of cognitive and neurodegenerative disorders, metabolic syndrome, type 2 diabetes, dyslipidemia or obesity.	US20190000867A1
Punicalagins	Functional food and beverage with increased antioxidant capacity for preventing or treating hypercholesterolemia and/or hypertension	EP2033526A1
Chestnut tannins	Antioxidant or anti-microbial additive, or agent for reducing nitrosamines or mycotoxins	EP2904910B1
Ellagitannins	Treatment of bacterial infections	US20110105421A1
GA, EA, isoquercitrin, tellimagrandin I and II, pedunculagin, TGGs, PGG and di-galloyl-hexahydroxydiphenoyl-D-glucose	Inhibition or prevention of obesity, lipid storage (reducing blood triglyceride levels), hyperlipemia, arteriosclerosis and thrombosis	US7687085B2
Gallotannins and ellagitannins	Regulation of the synthesis and secretion of cytokines, including TNF-α and IL-1β	US20080070850A1
Ellagitannins	Anti-inflammatory or anti-allergic agent by the inhibition of histamine release from mast cells. Regular oral administration of product can ameliorate or prevent rhinitis, atopic dermatitis or asthma	EP0727218A3
1,3,4-tri-galloylquinic acid, galloylshikimic acid derivatives strictinin, corilagin, castalagin, vescalagin, chebulinic acid, chebulagic acid, punicalin, punicalagin, punicacortein C, cannamtannin B2	Inhibition of the propagation in human cells of a human retrovirus (HIV)	CA2001898A1
Tellimagrandin	Inhibition of Gram-positive bacteria (*Staphylococcus aureus*) growth, anti-inflammation and leukemia treatment	US8975234B2

GA: gallic acid; EA: ellagic acid; PGG: pentagalloylglucose; TGG: trigalloylglucose; AP: aerial parts, F: flowers, L: leaves, P: petals, R: roots, S: seeds, St: stems. ns: not specified.

1.2. Circular Economy and Exploitation of By-Products

In this context, the wide variety of biological activities exerted by tannins and their natural sources are the perfect scenario for the implementation of a valorization strategy. In recent decades, natural products and ingredients have gained an increasing demand instead of the use of synthetic additives. Consumers opt for this option as they are safer, ecofriendly and they show plenty of health benefits, while avoiding side effects associated with synthetic antimicrobials [21,22]. Additionally, the concept of a circular economy has been boosted in recent years, whose principal idea relies on closing loops, creating complete cycles of production [23]. Several studies have suggested that the food production system could join the circular economy model by adapting its manufacture model and valorizing by-products of the agri-food industry [24]. Hence, considering that tannins are widely distributed among vegetation of the terrestrial and aquatic environments and the disposal of by-products of the agri-food industry such as leaves, peels or seeds that can be used as tannin-rich sources, valorization of tannin recovery might be a feasible approach [2,25,26]. Besides, brown algae are also potential sources of tannins as they are easily harvested, sometimes underutilized and, in other cases, considered as invasive species and, thus, their elimination from the environment is advisable [27]. The presence of tannins in agricultural wastes opens the possibility to obtain them from sustainable and affordable sources. This sustainable approach must further be supported by the continuous search for and optimization of novel extraction methods (i.e., solid liquid, ultrasound, microwave, supercritical fluids or high-pressure extraction) together with the use of green solvents [2,28]. Then, obtaining added-value products from underutilized by-products contributes to the development of a circular economy. Altogether, these measures would contribute to lowering the environmental impact of human activity, generate new and affordable added-value products and reduce the economic cost of recycling and waste management [29]. Yet, waste derived from tannin-containing plants can be used to extract these valuable products, with multiple potential applications [30].

Considering the aforementioned properties of tannins and their availability in nature, this review article is aimed at compiling the main biological activities of tannin-rich extracts to evaluate their potential use for food, nutraceutical and pharmaceutic applications, valorizing by-products of the agri-food industry as potential sources to produce added-value tannin-based products.

2. Biological Activities of Tannin-Rich Extracts

As aforementioned, tannins represent a chemical defense barrier for plants and algae that improve the response against pathological attacks and adverse abiotic conditions. The biological activity in plants and algae has prompted their utilization as traditional remedies to treat numerous diseases or infections. Currently, the biological effects of purified tannins or tannin-rich extracts (containing additional biomolecules) have been evaluated in vitro and in vivo using animal models, and more recently by clinical trials performed on humans [31]. Most of these research works have been focused on the study of the bioactivities of plants containing high amounts of tannins or, less commonly, purified tannins, to disclose their potential for developing innovative applications in the field of medicine, pharmacology, cosmetics, botany and/or veterinary medicine [10]. Among the biological activities of tannins, the most relevant ones are antioxidant, anti-inflammatory, anti-diabetic, cardioprotective, healing and antimicrobial (antiviral and antibacterial) [10,32] (Table 2).

Table 2. Tannin-rich genera with some representatives and their tannin chemical profile, including major compounds and their reported bioactivities.

Source	Species	Classification	Compounds	Bioactivities	Ref.
Acacia sp.	*A. mearnsii*	CT	Epi-FIS derivatives	Antioxidant, anti-inflammatory, antimicrobial	[33,34]
	A. nilotica	CT	PoGG, EA, GA, diGA, epi/gallocatechin, dicatechin derivatives	Antinociceptive, anti-inflammatory and antipyretic	[35,36]
Castanea sp.	*C. sativa*	HT	CAST, VES, EA, chestanin	Antioxidant, anti-inflammatory, antidiabetic, cardioprotective, antimicrobial, antifungal, antidiarrheal (vet.)	[37–45]
Juglans sp.	*J. regia*	HT	EA, pedunculagin, casuariin	Antiplatelet, cardioprotective, antiatherogenic and anti-inflammatory	[46–48]
Lotus sp.	*L. corniculatus* *L. pedunculatus*	CT CT	Heteropolymers PC: PD	Improvement of animal performance	[49–51]
Picea sp.	*P. abies*	CT	-	Antioxidant (food preservative)	[52]
Punica sp.	*P. granatum*	HT	Punicalagin, punicalin, geraniin	Antiviral (herpes simplex-2, hepatitis B)	[53,54]
Quercus sp.	*Q. robur*	HT	Castalin, vescalin, CAST, VES, GA, EA, PoGG	Antioxidant, antidiabetic	[55–57]
Rhus sp.	*R. coriaria*	CT HT	GA, QUERG, CYANG derivatives	Antimicrobial, anti-inflammatory, immunomodulatory, antiapoptotic and healing	[58,59]
Rubus sp	*R. fruticosus*	CT	CYANG, GA, malvidin-3-galactoside, vanillic acid	Antioxidant, anti-inflammatory, antidiabetic and gastroprotective	[60,61]
Sargassum sp.	*S. fusiforme*	PT	Eckol, dieckol, fuhalols	Antioxidant	[62]
	S. muticum	PT	PG, diphlorethol, bi- and tri-fuhalol A, B	Antioxidant, antibacterial, antiproliferative, anti-inflammatory	[63]
Schinopsis sp.	*S. lorentzii*	CT HT	FIS-catechin polymers TGG, PGG, quinic acid-GA esters	Antioxidant, antimicrobial, anthelmintic	[64–67]
	S. balansae	CT	ProFIS polymers	Antioxidant, antimicrobial, anthelmintic	[1,64–108]
Terminalia sp.	*T. chebula*	HT	Chebulinic acid, TGG	Anti-inflammatory	[68]
Vitis sp.	*V. vinifera*	CT	Galloylated PC, PC, PD	Antioxidant, anti-inflammatory, antiobesity	[69]

Definitions: CAST: castalagin, CT: condensed tannin, CYANG: cyanidin-3-glucoside, EA: ellagic acid, FIS: fisetinidin, GA: gallic acid, GT: gallotannin, HT: hydrolysable tannin, PC: procyanidin, PD: prodelphinidin, PG: phloroglucinol; PGG: pentagalloylglucose, PoGG: polygalloylglucose, PT: phlorotannin, QUERG: quercetin-3-glucoside, TGG: trigalloylglucose, VES: vescalagin.

2.1. Antioxidant

Tannins, as other polyphenols, present the ability to scavenge diverse free radicals and inhibit lipid peroxidation. In fact, their content increases under stressful conditions in cellular pro-oxidant states. This activity is related to the presence of phenolic rings in the chemical structure of the compound and also the degree of polymerization [2,70].

The application of tannins as antioxidants has been evaluated in living cells. For example, tannic acid is commonly utilized as purified tannin. Its antioxidant capacity has been demonstrated in an in vitro assay based on fibroblasts irradiated with UVB to create an oxidative environment that led to cellular damage, simulating photoaging. Tannic acid showed strong antioxidant properties, a broad UV absorption spectrum and also an inhibitory activity towards collagenase and elastase. Thus, this tannin acid was demonstrated to prevent photodamage by attenuating the evaluated oxidation levels and diminished photoaging parameters [71]. The antioxidant activity of many other types of tannins has been evaluated. In a previous study, the ability to prevent lipid peroxidation of diverse phenolic compounds, including 25 tannins (both CTs and HTs at 5 μg/L), was assessed in rat liver mitochondria. The results displayed that HTs, like pedunculagin, PGG and chebulinic acid, were the most effective inhibitors. This suggest that the presence of structures such a galloyl, HHDP or dehydro-HHDP are involved in the inhibition of lipid peroxidation [72]. Similar results were observed when different tannins (CTs: catechin and epigallocatechin-gallate; HTs: PGG and geraniin) were used to inhibit the effects of the induced lipid peroxidation in mouse lens [73]. The seed coat from *Phaseolus vulgaris* was used to extract and purify tannins and flavonoids and the antioxidant activity of these compounds was tested using a liposome assay. Delphinidin and petunidin-3-glucoside were the most active tannins, showing an activity close to the 50% of that of the synthetic antioxidant butylated hydroxytoluene (BHT) [74]. Another work characterized a CT form from *Diospyros kaki* and then analyzed its antioxidant capacity by an ex vivo tissue system and an in vivo assay. When tested in a mouse liver homogenate, it showed strong protection against auto-oxidation and H_2O_2-induced oxidation processes with half inhibition concentrations of 4.3 and 1.4 μg/mL, respectively. In vivo assays, performed by the oral administration of 200 or 400 mg of CT per kg of mouse body weight, displayed a reduction of the activities of the evaluated oxidative biomarkers (serum and liver superoxide dismutase (SOD), GSH (reduced glutathione) peroxidase and liver malondialdehyde (MAD) activities) [75]. The antioxidant properties of tannin-rich extracts obtained from different vegetal species have been assessed. For instance, a *Q. robur* tannin-rich extract, fundamentally composed of roburins, castalagin and vescalagin, has been reported to ameliorate oxidative stress markers and serum levels of related enzymes such as catalase (CAT) or SOD in a clinical trial [56]. Another trial with the same extract studied the effect in vivo and ex vivo, analyzing the plasmatic oxidative profile and genetic expression of cell cycle-related genes from plasma cell samples and several tissues. Although the study sample consisted of only three subjects, the results were significant and very similar in all three subjects, with a significant increase in phenolic concentration in plasma, as well as the modulation of targeted genes [57]. A study evaluated an extract from *A. mearnsii*, which displayed antioxidant properties that reversed the negative effects caused by acrolein (a compound related to neurodegenerative diseases)-induced cytotoxicity in a human neuroblastoma cell line (SH-SY5Y). In addition, the extract also inhibited the action of apoptotic factors [76]. In the same line of experiments, oxidative stress was induced in the SH-SY5Y cell line after its previous treatment with *C. sativa* extracts, which significantly reduced reactive oxygen species (ROS) production. In addition, it was observed that the previous treatment reduced apoptotic signals caused by the damage inducers [37].

Although the antioxidant mechanism of tannins has been repeatedly investigated, deeper studies about the concrete mechanism of action are needed, especially considering the administration as well as the variability of the tannin metabolic profile associated with each species. Additionally, it is worth mentioning that the antioxidant capacity is the basis for triggering further systematic and beneficial effects, such as anti-inflammatory responses and wound healing (Figure 2).

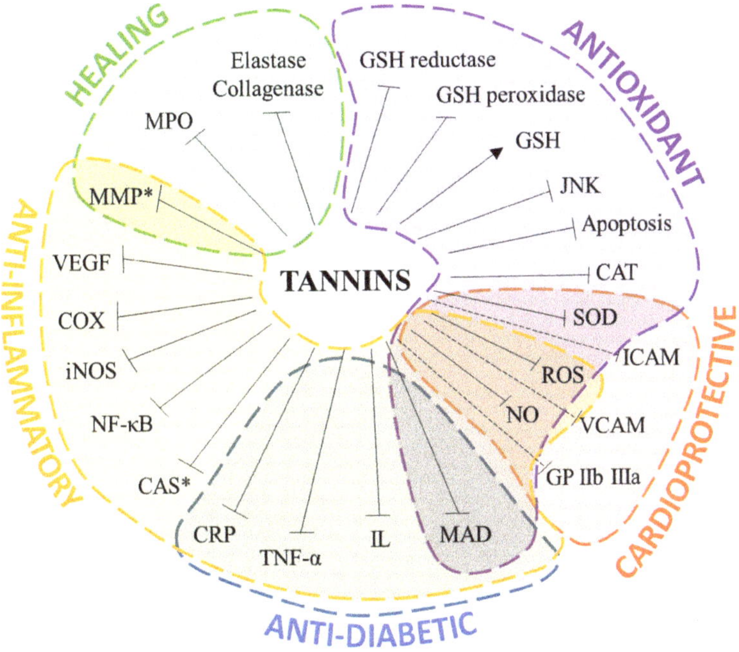

Figure 2. Visual representation of the suggested mechanisms involved in the biological properties of tannins. Lines show decrease in or inhibition of biomarkers, whereas arrows show an increase in or promotion of reduced glutathione (GSH). (* = antioxidant activity; MAD: malondialdehyde; IL: interleukin; TNF-α: tumor necrosis factor-α; CRP: c-reactive protein; CAS: caspase; NF-κB: nuclear factor-κB; iNOS: nitric oxide synthase; COX: cyclooxygenase; VEGF: vascular endothelial growth factor; MMP: matrix metalloproteinase; JNK: C-Jun N-terminal kinase; MPO: myeloperoxidase; CAT: catalase; SOD: superoxide dismutase; ROS: reactive oxygen species; NO: nitric oxide; VCAM: vascular cell adhesion protein; ICAM: intercellular adhesion molecule; GP IIb IIIa: glycoprotein IIb/IIIa).

2.2. *Anti-Inflammatory*

Recently, numerous works have disclosed the mechanism of action of the anti-inflammatory effect of several tannins. However, many other works demonstrate the systemic effects of these natural molecules without presenting the specific cellular mechanism or without identifying the specific compounds responsible for the effect. We have tried to focus on those providing the chemical profile and mechanism of action.

An in vivo assay using mice applied an aqueous extract from the bark of *A. nilotica* by intraperitoneal injection to determine its antinociceptive, anti-inflammatory and antipyretic activity. The extract results displayed a slight reduction in paw edema. In addition, *Acacia* treatment (150 mg/kg body weight) was able to inhibit the formalin-induced inflammation at values like those of diclofenac sodium. The antipyretic treatment was maintained for 3 h and showed a slight inhibitory effect on fever after inducing the pyrexia with yeast [36]. Another study, using in vitro techniques, analyzed different extracts from leaves of *A. mearnsii*. The most active one possessed at least (epi)fisetinidol derivatives (with catechin, gallocatechin, additional molecules of fisetinidol or even robinetinidol) quantified at 12.6 mg/g as procyanidin B2 equivalents. This fraction, applied at non-cytotoxic levels (50 μg/mL) in RAW 264.7 macrophages, previously exposed to oxidative stress, was able to significantly inhibit ROS production and reduce nitric oxide

(NO) back to non-stimulated levels. Later, the same cell culture was exposed to an inflammatory process through lipopolysaccharide (LPS) stimulation. This assay showed that the *A. mearnsii* extract inhibited the expression of cytokines like interleukin-1β or -6 (IL-1β, IL-6) and pro-inflammatory enzymes such as cyclooxygenase-2 (COX-2) or inducible nitric oxide synthase (iNOS) [33]. Similar results were observed in LPS-stimulated RAW 264.7 macrophages treated with a *T. chebula* extract. Among the identified molecules, two GTs (chebulinic acid and 2,3,6-tri-O-galloyl-β-D-glucose) applied at 50 μM could reduce NO production and decreasing the protein expression of iNOS and COX-2 [68]. The anti-inflammatory properties of extracts from the spiny burs of *C. sativa* have also been tested in vitro using LPS induction in the BV-2 cell line, simulating a microglia model. The treatment showed cytoprotection of the downregulation of the expression of IL-1β, tumor necrosis factor-α (TNF-α) and nuclear factor-κB (NF-κB) [38]. Following similar approaches, an inflammatory process was induced in the HaCaT cell line using tumor necrosis factor-α (TNF-α). Then, the cells were treated with ethanolic extracts from *R. coriaria* obtained by maceration or cold extraction. The induction with TNF-α stimulates pro-inflammatory signals by the production of interleukins, vascular endothelial growth factor (VEGF), matrix metallopeptidase 9 (MMP-9) and intercellular adhesion molecule 1 (ICAM-1). This inflammatory cascade was inhibited by both kinds of *R. coriaria* extracts, except for VEGF, which was just decreased by the maceration extract [58]. Another study, performed in vivo, orally administrated *R. coriaria* extracts to rats to study its ability to prevent or treat necrotizing enterocolitis. The antioxidant, anti-inflammatory, immunomodulatory and antiapoptotic abilities of *R. coriaria* were analyzed through the quantification of oxidative indicators and histological assays. The application of the treatment reduced the presence of inflammatory molecules in histological samples, while biochemical results reported lower amounts of IL-6, TNF-α and lipid hyperoxides. Besides, the negative effects of induced necrotizing enterocolitis were reversed [59]. In another in vivo study using rats, the ability of procyanidins obtained from grape seeds to reduce inflammation induced by a hyperlipidic diet was analyzed. The oral administration of these tannins produced a down-regulation of C-reactive protein (CRP), TNF-α and IL-6 in liver and white adipose tissue [77]. However, in a comparative work of extracts from *R. occidentalis* and *Vitis labrusca* seeds, the analysis of the former showed higher contents of tannins and also stronger antioxidant and anti-inflammatory properties [78]. Indeed, many works have been carried out using, as a basis, species belonging to the genus *Rubus* to show their potential bioactivities. For instance, an extract of *R. fruticosus* was evaluated as an antioxidant, anti-inflammatory and gastroprotective agent in rats. The anti-inflammatory effects reported by the histological exam were attributed to cyanidin-3-glucoside through the reduction or inhibition of the activity of NF-κB, COX-1 and -2, NO and/or iNOS [60]. In vitro assays using another species of the same genus, *R. idaeus*, demonstrated the reductio of inflammation and oxidation in hypertrophied adipocytes. Extracts from fruits of *R. idaeus* were able to down-regulate the expression of IL-1β and -6, TNF-α and leptin but also up-regulate the expression of antioxidant enzymes, such as SOD and CAT. Apart from these main mechanisms, the application of *R. idaeus* extracts reduced lipid accumulation and increased lipid mobilization in hypertrophied adipocytes, which may help to prevent the future appearance of further metabolic disorders [79].

As shown in these previous works, antioxidant and anti-inflammatory activities of tannins can have positive collateral effects. Therefore, tannins have been tested as natural ingredients with preventive or treatment purposes in many diseases or infections whose main bases are oxidative and inflammatory processes such as diabetes, heart infections or wound healing.

2.3. Antidiabetic

In a recent in vivo experiment performed in rats, the efficacy of *R. fruticosus* as a source of natural antidiabetic agents was supported. hydroethanolic extracts of *R. fruti-*

cosus were administered by intraperitoneal injection to streptozotocin-induced diabetic rats. Diabetes, like many other chronic diseases, has been found to trigger oxidative and inflammatory processes at a cellular level. The intraperitoneal administration of *R. fruticosus* was demonstrated to reduce oxidative and inflammatory markers such as TNF-α, IL-6 and CRP [61]. Different species recognized as tannin-rich sources have been also described as potential sources of antidiabetic agents, such as *C. sativa*. *Q. robur, S. lorentzii* or *T. chebula*. Tannins, especially HTs, have been reported to inhibit α-glucosidase, an enzyme responsible for the absorption of carbohydrates from the gut. In fact, inhibitors of α-glucosidase may be used in the treatment of patients with type 2 diabetes mellitus (DM) or impaired glucose tolerance [39]. Therefore, the administration of tannins for the prevention or treatment of type 2 DM may have a doubly positive effect, as an antioxidant and as an α-glucosidase inhibitor. For instance, different extracts from the wood of *C. sativa* were tested as potential α-glucosidase inhibitors and as antioxidants. An initial extract was fractioned into five parts, from which the best one was fractioned into seven parts. From these seven fractions, the ones with the strongest antioxidant capacity were mostly composed of the phenolic acid GA, grandinin, valoneic acid dilactone and its galloyl derivative and trigalloylglucose (TGG) molecules. The extracts with stronger α-glucosidase inhibition contained valoneic acid dilactone, three TGG isomers and PGG. The molecule common to these two extracts was the valoneic acid dilactone that has previously been reported as an α-glucosidase inhibitor [39]. Similarly, from *Q. robur*, different fractions were tested for antioxidant and α-glucosidase inhibition activities, with the molecules involved in the strongest antioxidant role being a monogalloylglucose (MGG) isomer, an HHDP-glucose isomer, castalin, GA, vescalagin and grandinin/roburin E isomer. The sub-fraction with the strongest α-glucosidase inhibitory activity contained castalagin as the major tannin [55]. From *S. lorentzii*, fractions composed of HTs, esters of quinic acid with different units of GA (di-, tri-, tetra- and penta-galloylquinic acids) and oligomeric CTs (dimers, trimers, tetramers or pentamers of catechin or catechin-3-O-gallate and fisetinidol, or catechin-3-O-gallate) possessed the strongest α-glucosidase and α-amylase activity [65]. The tannin associated with the antidiabetic activity in *T. chebula* was corilagin [80]. Other components of *T. chebula*, like chebulanin, chebulagic acid and chebulinic acid, have been demonstrated to be able to inhibit the activity of maltase, an enzyme with a high activity rate in diabetic processes [81]. Furthermore, ETs and GTs isolated from *T. bellerica* and *T. chebula* have been described to improve the peroxisome proliferator-activated receptor-α and/or -γ signaling, which plays an important role in controlling the expression of genes related to the storage and mobilization of lipids, glucose metabolism, morphogenesis and inflammatory response, which have direct effects in insulin sensitivity [82].

2.4. Cardioprotection and Blood Circulation Improvement

The potential of polyphenols as natural antiplatelet, anti-inflammatory, and anticoagulant agents has prompted their analysis from a cardioprotective point of view. In this sense, numerous studies have demonstrated the beneficial cardiac effects of the walnut (*J. regia*). The major phenolic compounds present in *J. regia* are ETs, including HHDP derivatives which are capable of releasing EA, a well-known antioxidant related to different health benefits [46]. A work based on human aorta endothelial cells analyzed the effects of an extract from peeled fruits of *J. regia*. Initially, inflammatory processes were induced in the cultured cells by their exposure to TNF-α, which prompted the maximal expression of vascular cell adhesion protein (VCAM-1) and ICAM-1. The co-treatment of cells with *J. regia* extracts or with purified EA, one of their major components, inhibited both inflammatory biomarkers [47]. Another experiment (in vivo) utilized isoproterenol, a synthetic catecholamine capable of producing myocardium pathologies, to induce myocardial infarction. This catecholamine was administered individually or in combination with *J. regia* extracts. The presence of walnut extracts reduced the severity of the myocardial infarction in a dose-dependent manner, quantified through serum creatine kinase myocardial band- levels and the activity of lactate dehydrogenase. Additionally, the administration of *J. regia* extracts was able to reverse the negative effects of isoproterenol on oxidative markers, myocardial tissue lipids and at a histopathological level [48]. Thus, these works, among many others, demonstrated the antiatherogenic and cardioprotective potential of the consumption of *J. regia*.

Other plants have also been evaluated in terms of cardioprotective activity. For instance, extracts obtained from leaves of *R. idaeus* were evaluated through a blood ADP assay to determine how they modify blood platelet aggregation. The results showed that *R. idaeus* reduced, by more than 20%, the expression of the glycoprotein IIb/IIIa, which is involved in the reception of fibrinogen and the activation of platelets. The activation of the aggregation was nearly inhibited to 50% by the presence of the extract. When the antiplatelet activity of the extracts was tested in blood, the inhibition of platelet aggregation was reduced to less than 20% [83]. Ethanolic extracts from seeds of *Acacia senegal* were tested as antiatherosclerotic and cardioprotective agents in an in vivo experiment with rabbits subjected to a hypercholesterolemic diet. The administration of the extract reduced the levels of the total cholesterol, low- and very low-density lipoprotein (LDL and VLDL) cholesterol and triglycerides in blood. Besides, *Acacia*-treated subjects showed a lower atherogenic index accompanied by a reversion in lipid oxidation markers and histological damage [84]. Finally, bark extracts from *C. sativa* were assayed using primary cultures of neonatal rat cardiomyocytes and cardiac tissues isolated from guinea pigs. Cardiomyocytes were exposed to H_2O_2 to induce oxidation states, while cardiac tissues were incubated with carbachol for testing the muscarinic activity, propranolol for the adrenergic/cholinergic activities and noradrenaline for evaluating aortic muscle behavior. *C. sativa*-treated cardiomyocytes showed a dose-dependent reduction of intracellular ROS production, which directly improved cell viability. The aortic noradrenaline-induced contraction was reduced by the *C. sativa* extract, which also reduced heart rate and produced a positive inotropic effect in the left atrium/papillary (adrenergic receptor involvement was demonstrated) and negative chronotropic effect (not mediated by cholinergic receptors). Thus, the results supported the use of *C. sativa* extracts as dietary supplements since they may provide synergic beneficial effects as cardioprotective and antioxidant agents [40].

2.5. Wound Healing

Tannins have been demonstrated to prevent the appearance of ulcers or to accelerate wound healing, which may have different further applications. That is the case of an experiment performed with *R. imperialis*, whose anti-inflammatory and wound healing activity was investigated. In vitro assays demonstrated the antioxidant and anti-inflammatory properties and lack of cytotoxicity of the extracts, while in vivo experiments focused on their cytoprotective and healing effects, also including anti-inflammatory responses. The administration of *R. imperialis* (100 mg/kg) via gavage was shown to block the migration of neutrophils, which had a positive effect in cutaneous wounds. Besides, in vitro assays using LPS to simulate inflammation and leukocyte migration showed that the application of *R. imperialis* increased fibroblast migration up to 76% when compared with the control and reduced NO release. Additionally, the topical administration of *R. imperialis* was shown to affect collagen proliferation with a better organizational pattern than the control. In addition, in vitro experiments displayed a very low hemolysis rate for the extracts and no skin irritation potential [85]. Another approach to the wound healing strategy was presented in a work where an extract of *R. coriaria* was studied for its potential healing, anti-inflammatory and antimicrobial activities. Anti-inflammatory markers, such as the activity of myeloperoxidase and matrix metalloproteinase-8 (MMP-8) enzymes, were reduced in animals treated with *R. coriaria*, while healing indicators, such as hydroxyproline or collagen deposition, increased and the wound area was reduced, completing epithelization and scar formation by the 10th day of treatment. Different in vivo and in vitro experiments supported the antimicrobial activity of the *R. coriaria* extract. Indeed, animals with wounds infected with *Staphylococcus aureus* or *Pseudomonas aeruginosa* showed a slight delay in the healing process, reaching similar markers between the 10th and 13th day of treatment [86]. Hence, as described in this work, the healing activity of tannins exerts not just a cicatrizing effect but also an antimicrobial effect, which is crucial since wounds often get infected. The antimicrobial capacity of tannins is reviewed in the next sub-section.

2.6. Antimicrobial

Many works have supported the antibacterial, antifungal and antiviral properties of tannins [10,31]. For example, walnut leaves have been approved for the topical treatment of mild skin inflammation, due to their anti-inflammatory and antimicrobial activities. An ethanolic extract of immature *J. regia* fruits was able to inhibit methicillin-resistant *S. aureus* (MRSA) growth when administrated in a concentration range of 128–512 µg/mL. At 16 µg/mL, the extract reduced biofilm formation and adherence. Thus, this extract was suggested to be an efficient treatment for skin and soft tissue infection processes [87]. Similarly, extracts obtained from *S. brasiliensis* have shown antimicrobial potential against MRSA [88]. A very recent work performed transcriptome and metabolome data analysis to find the mechanism of action of tannins from *Diospyros kaki* against MRSA. The results suggest that some main mechanisms are related to physical damage of the bacterial cell membrane [89]. Tannins obtained from *A. mearnsii* were evaluated as antibacterial agents, both as free and encapsulated molecules. Free tannins showed antibacterial activity, especially against *S. aureus*, with MBC (Minimum Bactericidal Concentration) of 0.32 and 1.25 of mg/mL. They also acted against fungal (*Aspergillus niger*, MIC (Minimum Inhibitory Concentration) 0.62 mg/mL) and yeast (*Candida* sp., MIC 2.5 mg/mL) growth. The most efficient antimicrobial encapsulated showed IZs for *S. aureus*, *Escherichia coli*, *A. niger* and *Candida* sp, which were larger of those observed with the free tannins, except for *S. aureus* [34].

Several species belonging to genus *Rubus* were tested in vitro as antibacterial and antifungal agents. The main tannins involved in these activities were different among the species and plant parts analyzed. Extracts were tested against two strains of *Helicobacter pylori*, with and without the chromosomal insertion *cag*, whose presence is associated with an increased inflammatory profile. The whole extract, after 24 and 48 h of treatment, showed values of 1200 and 134 µg/mL of minimum bactericidal concentration (MBC) against the cag- *H. pylori*. Tannins and other phenolic compounds have been suggested to be involved in the antibacterial mechanism by the inhibition of bacterial ionic pumps [90]. A species from the same genus, *R. ulmifolius*, has been evaluated in terms of amoebicidal, antibacterial and antifungal activities. Trophozoites from *Acanthamoeba castellanii* showed a dose-dependent sensitivity to the extract, but not comparable with the positive control. The antibacterial and antifungal capacity of extracts was determined by inhibition zone (IZ) diameters, minimal inhibitory concentration (MIC) and MBC. *R. ulmifolius* extracts presented MIC and MBC for all bacteria and the yeast in the range of mg/mL. The best results in terms of IZ were achieved for *Escherichia coli*, *Streptococcus agalactiae* and *Candida albicans*, while Gram-positive *S. aureus* and *Enterococcus feacium* were less sensitive [91]. Furthermore, purified methanolic extracts of *R. ulmifolius*, which presented a high content of tannins (both HTs and CTs) demonstrated significant antifungal activity against five filamentous fungi: *Beauveria* sp., *Fusarium solani*, *Microsporum canis*, *Phialophora verrucosa* and *Scopulariopsis brevicaulis* [92].

Different parts of *C. sativa* (leaves, burs, outer and inner shells) were also tested for antibacterial activity. Leaves contained the highest phenolic composition, with trigalloyl-HHDP-glucose-like molecules, as major tannin representatives, also present in burs. Outer shells were rich in GA, while the main compound in the inner shell was syringetin-hexoside. Extracts from inner shells were able to inhibit the growth of the Gram-positive bacteria *Staphylococcus epidermidis*, *S. aureus*, *Enterococcus faecalis* and *E. faecium*, and Gram-negative *Klebsiella pneumoniaea* and *P. aeruginosa*, with MIC values between 25 and 50 mg/mL [41]. Other tannins present in *C. sativa*, such as vescalagin and castalagin, were isolated, purified and demonstrated to have antibacterial activity against *E. coli*. Apart from them, commercial crude extracts obtained from quebracho, chestnut and mimosa, and two classes of tannic acid and one of GA, were analyzed. From the results, it was observed that tannic acid possesses much better growth inhibitor activity than GA against *E. coli*. On the other hand, a crude extract of chestnut had stronger antibacterial properties than purified vescalagin and castalagin, probably due to the synergy exerted by the molecules contained in the extract [42]. Finally, *C. sativa* extracts have been reported to inhibit *E. coli* and *Clostridium perfringens* growth when applied at 1200 μg/mL and 3–150 μg/mL, respectively [43,93].

Tannins extracted from different genera have also been experimentally demonstrated to act as antiviral agents. For instance, an extract of *P. granatum*, with punicalagin, GA and EA as major components, was analyzed against herpes simplex virus 2. The compound punicalagin showed significant antiviral activity, comparable with the positive control. However, when used as part of the extract, the required concentration to achieve total inhibition was higher [54]. Apart from punicalagin, other work analyzed the antiviral activity of punicalin and geraniin against hepatitis B virus (HBV). When these three HTs were tested in a human hepatocyte cell line (HepG2.117), they showed a dose-dependent reduction of supernatant e antigen levels, which indicated that the tannins were interfering with the synthesis, stability or transcription of the viral DNA [53]. *Urtica dioica* and *Taraxcum officinale* exhibited inhibitory effects in the range of 126–166 μg/mL against dengue virus serotype 2 when tested in vitro using hamster kidney cells (BHK-21). Recently, an in silico evaluation of the application of 19 HTs was performed to screen their potential ability to inhibit the activity of SARS-CoV-2. Specifically, the potential allosteric ligand of different HTs with 3-chymotrypsin-like cysteine protease enzyme (described to be involved in virus transcription) was evaluated. Among the tested HTs, pedunculagin, tercatain and castalin interacted with the catalytic dyad Cys145 and His41 through stronger binding forces. Other HTs, like tellimagradin I, punicalin, chebulagic acid or β-pedunculagin, may have secondary roles in the inhibition of the activity of this catalytic target [94].

2.7. Other Beneficial Applications of Tannins

2.7.1. Human Beings

Nowadays, there is plenty of evidence for the ethnopharmacological use of tannin-rich plants as antidiarrheal treatments. Yet, some clinical studies employing tannin extracts have shown promising results regarding the efficacy of their use and their safety. A study described how the administration of tannins reduced the duration of diarrhea caused by rotavirus in infants. Similarly, a more recent study reported that children affected by acute diarrhea presented a significant decrease in the duration of the diarrheal symptoms when administered with tannins in comparison with the standardized rehydration treatment [95,96]. These findings not only support the use of tannins as effective antidiarrheal treatments, but also provide information on their safety, since the test subjects were children. In the same way, a recent study with quebracho and chestnut extracts studied the antioxidant activity and metabolization of these extracted tannins in in vitro digestion–fermentation assays. The results evidenced the degradation of tannins by gut microbiota, producing metabolites like quercetin or sinapinic acid, as well as higher antioxidant capacity on residual solids after fermentation and increased production of short-chain fatty acids. Short-chain fatty acids are described as prebiotics in the gut [97]. This would infer that tannins also have a prebiotic effect on digestive microbiota.

A clinical trial with patients prone to developing urinary tract infections unveiled the potential of tannins as a treatment against these infections. In that case, tannins were extracted from *Serenoa repens* and orally administered as a food supplement. Although the results differed in a gender-dependent manner, leukocyte count and urinary microbiota decreased significantly in the subjects after 9 weeks, which is a relevant result, given that these patients tend to show higher levels of these markers [98].

Another uncharted potential application of tannins may be their use as anxiolytics. A recent study analyzed the performance of *T. chebula* tannin extracts on anxiety behavior, the genetic expression of gamma-aminobutyric acid (GABA) receptors and corticosterone markers, as well as in electroencephalogram assays in mice. The results appear to indicate that tannin extracts were able to ameliorate the expression of GABA receptors, selected biochemical markers and improve the oxidative status of cerebral tissue [99].

2.7.2. Veterinarians

A great number of studies have evaluated the potential use of tannins and tannin supplements in livestock, such as cattle, poultry or sheep. Research has focused on their use as antimicrobials or growth promoters. Nonetheless, tannins are known antiherbivore compounds able to reduce the digestibility of proteins in said animals by their aggregation properties [31].

Legumes have been suggested as an additional ingredient to create pasture forages since they represent a protein source for livestock, improve the rumen microbiome and reduce greenhouse gas emissions. Among the range of legumes, *Lespedeza procumbens*, *Desmodium paniculatum*, *Leucaena leucocephala*, *D. ovalifolium* and *Flemingia macrophylla* possess a relatively high content of CTs. They have been suggested to modify rumen fermentation in beef cattle, probably due to the presence of tannins, so they may act similarly in animals by complexing proteins and short-chain fatty acids, which may provide a prebiotic effect, as stated [100]. Tannins have exhibited similar antidiarrheal results in cattle as those mentioned in humans. As an example, a blinded study tested the effect of a proprietary chestnut tannin extract on the duration of neonatal diarrhea in calves. The results were consistent with other previous data, notably a lowering of the duration of the diarrheic episodes without affecting the weight of the animals [45]. CTs from *L. pedunculatus* and *L. corniculatus* are reported to avoid protein degradation and to improve amino acid residue absorption at the intestinal level, which ultimately enhances animal performance, with an apparent higher ovulation rate and clean wool production in sheep [51]. Finally, tannins have also been employed to treat gastrointestinal diseases. The anthelmintic activity of tannins has been addressed, but they are also effective against protozoan parasites. This is the case of a study where *Berberis vulgaris* and *R. coriaria* extracts were applied against relevant pathogenic protozoans like *Theileria equis* and common *Babesia* species, such as *B. bovis*, *B. bigemina* or *B. caballi*, in a mouse-infected model. The experiment showed better levels of selected biomarkers in plasma, due to the synergetic effect of compounds present in the used extracts, of which tannins and flavonoids were detected as the main chemicals [101].

2.7.3. Botanical

Tannins have been subjected to further research to evaluate their potential use as antifungal compounds to treat or enhance plant resistance against plant pathogens. Yet, most studies have been performed in vitro and knowledge gaps on the ecological role of tannins remain. *P. granatum* peel extracts have been tested both in vitro and in vivo against *Pseudomonas syringae*, a common and hazardous pathogen of tomato. The results demonstrated that bacterial growth was inhibited for up to 15 days after leaf inoculation [102]. In vitro antifungal assays carried out with chestnut (*C. sativa*) bur extracts, mainly composed of GA and EA moieties, as well as ETs and glycosylated flavonols, inhibited the growth and spore germination of common plant-infectious fungi *Alternaria alternata*, *Botrytis cinerea* and *Fusarium solani* [44].

2.7.4. Food Additives

As previously shown, tannins have been demonstrated to be inhibitors of lipid peroxidation and scavengers of free radicals, which are the main reason for the appearance of rancid off-flavors/aromas in foods. Therefore, the utilization of tannins as a natural antioxidant in food applications is drawing attention. Among the purified tannins, the most tested is tannic acid, which has been demonstrated to reduce lipid oxidation induced by ferrous ions in a plant-based emulsion of flaxseed oil droplets. In this assay, the antioxidant activity of the tannic acid was attributed to the metal binding properties of tannins [103]. The same molecule was applied to ground chicken breast meat to determine its ability to reduce lipid and protein oxidation, maintain color and prevent rancid volatiles. The addition of 5 to 10 ppm of tannic acid improved both cooked and raw quality parameters, showing low oxidation markers, off-odor volatiles and high color parameters [104]. Extracts of tannin-rich plants such as quebracho (*S. balansae* and *S. lorentzii*) or conifers (*P. abies* or *Pinus sylvestris* L.) have also been tested as natural food preservatives with antioxidant properties to improve the shelf life of different products. Quebracho tannins (0.5–1.5%) were applied to beef patties, showing that the lowest amount was able to improve lipid stability. On the other hand, higher concentrations reduced the tenderness, softness and juiciness of meat [66]. Another study analyzed conifer tannins as antioxidants in a liposome model and in meat snacks. The tested tannins showed a high activity to prevent lipid oxidation without causing organoleptic interference in meat snacks [52]. Altogether, this research suggests that tannins, as for other polyphenols, could be added to food matrices as alternative antioxidants.

3. Valorization Approach and Concluding Remarks

As was mentioned before, interest in a circular economy and food or agricultural waste valorization is increasing. Therefore, biorefinery approaches for recovering bioactive molecules with target biological properties are consequently growing to face the current challenge: moving towards a circular system production model [24,105]. The use of by-products from the agri-food industry for the recovery of tannins was proposed decades ago. In 1990, it had been already suggested that pods, seeds, cake or kernel residues of some products, such as mango or cocoa, could serve as potential sources of tannins [106]. Their presence has been assessed in different by-products of the agri-food industry, such as green tea processing residues and acorn, chestnut and persimmon hulls and starch, showing antioxidant or antimicrobial properties, among others [107]. More recently, other studies have also addressed the possibility of recovering tannins from a secondary residue, such as distilled waste by-products remaining after the steam distillation of the underutilized biomass of specific trees [108]. Additionally, other experiments have been focused on assessing the bioavailability of tannins and the cellular level where they are located, i.e., they were found inside the cell lumen of parenchymatic cells and the vessels of chestnut wood [109]. Further applications are focused on obtaining other products from tannin-based sources. For instance, coffee pulp (rich in tannins) was submitted to solid state fermentation by *Penicillium verrucosum* to produce tannase [110]. However, to date, one of the biggest concerns regarding obtaining tannins is their high diversity, since this is related to their origin, extraction and purification procedures [111]. Hence, conventional and novel extraction techniques have been investigated to optimize the extraction parameters of tannin recovery from different by-products (Table 3).

Table 3. Examples of tannin extraction techniques from different agri-food by-products.

Species	Tannin	By-Product	Extraction Techniques	Experimental Conditions	Activity	Ref.
Trapa quadrispinosa	HT	Pericarps	UAE	Et/W (60/40, *v*/*v*), 30 min, 40 °C, L/S ratio 40 mL/g	Antioxidant (DPPH)	[112]
Cupressus lusitanica and *Cistus ladanife*	TTC	Waste distilled after steam distillation	UAE	Et S/L ratio 1:20, 30 min, 30 °C, 70% A	Antioxidant (ABTS)	[108]
Coffee (*Coffea arabica*)	Procyanidins (CT)	Pulp	UAE	W/A extract, 20 min, RT	-	[113]
Pomegranate (*P. granatum* var. *Gabsi*)	TTC	Peels	UAE	2.63 g/100g dw, 55.46% E, 30 min	Antioxidant (DPPH and ABTS)	[114]

Table 3. *Cont.*

Species	Tannin	By-Product	Extraction Techniques	Experimental Conditions	Activity	Ref.
Red grape variety (*Vitis vinifera*	CT	Pomace	HAE	NaOH, Na_2CO_3 or $NaHCO_3$) and Na_2SO_3 (2.5% or 5% (*w/w*). S/L ratio 1:8, 120 min, 100 °C	Production of environmentally friendly wood adhesive	[115]
					Silver NPs, antimicrobial and apoptotic potential	[116]
Chestnut (*Castanea sativa*)	TTC	Shells	Maceration	Et (20 mL × 3 days × 3 times) or Et/W 7:3 *v/v* (20 mL × 3 days × 3 times)	Antioxidant (DPPH and TEAC)	[117]
Pomegranate (*Punica granatum* L.)	TTC	Peels	HAE	W, 2% SS and 0.5% SB, S/L ratio 1:5, 7 h, 80 ± 5 °C	-	[118]
Tea (*Camellia sinensis* L.)	TTC	Leaves	SFC-CO_2	Supercritical CO_2 flow rate 8 g/min, 188 bar, 50 °C, co-solvent flow rate 2.94 g/min	Antioxidant (ABTS)	[119]
Acacia mollissima	HT and CT	Bark	HAE and MAE	HAE: M (2h, 20 °C and 60 °C). MAE (1 min, 300W or 5 min, 150W)	-	[120]
Myrtus communis L.	TTC	Leaves	MAE	Et 42% (60 s, 500 W, S/L ratio 32 mL/g)	Antioxidant (DPPH, TEAC and ORAC)	[121]
Endopleura uchi	TTC	Bark	Maceration	Et/W 50%	Antimicrobial, cytotoxic and antioxidant	[122]
Norway spruce (*Picea abies*)	CT	Bark	Hot water extraction	10% solid content, 2% SS, 0.5% SC, (75 °C, 120 min)	-	[123]
Spruce (*Picea abies*)	TTC	Bark	SFC-CO_2	Solvent consumption 2.5 kg CO_2/kg product and 24.94 kg Et 70/kg product, 100 bar, 40 °C	Antioxidant (DPPH)	[124]
Eucalyptus globulus	EA and GA	Leaves	BMSHE	1.0 M [$HO_3S(CH_2)$4mim] HSO_4, L/S ratio 30 mL/g. MAE: 20 min, 385 W)	-	[125]

Definitions: TTC: total tannin content, Et: ethanol, W: water, A: acetone, M: methanol, SS: sodium sulfite, SB: sodium bicarbonate, SC: sodium carbonate, RT: room temperature, DPPH: 2,2-diphenyl-1-picrylhydrazyl, TEAC: trolox equivalent antioxidant capacity, ORAC: oxygen radical absorbance capacity, ABTS: 2,2′-azino-bis(3-ethylbenzothiazoline-6-sulfonic acid, UAE: ultrasound-assisted extraction, HAE: heat-assisted extraction, MAE: microwave-assisted extraction, SFC-CO_2: supercritical carbon dioxide extraction, BMSHE: Brønsted acidic ionic liquid-based microwave-assisted simultaneous hydrolysis and extraction.

Taken all together, tannins are phenolic compounds that have been used in traditional medicine, are widely distributed and have been broadly investigated for their biological properties. They can be classified according their structure and obtained mainly from vegetal sources and marine brown algae. Specific tannins of some genera or species, such as

HTs, castalagin, vescalagin or punicalagin, seem to have relevance for biological activities. Important specificity differences were also observed for compounds belonging to ETs or GTs, due to the phenolic acids released after their hydrolysis (EA or GA, respectively). The most remarkable biological activities of tannins have been recorded as antioxidant, anti-inflammatory, antidiabetic, cardioprotective, healing and antimicrobial. Nevertheless, many of the indicated activities act synergically, with the antioxidant capacity being the main axis of the connection between the different biological properties. Indeed, all these interactions among biological activities are easily demonstrable, since common molecular targets are described to be involved in the different biological activities. Therefore, considering all the biological properties described in tannins obtained from natural sources, valorization could be an efficient approach to revalorize agri-food by-products. However, further study is still necessary to completely elucidate the mechanisms of action of the biological activities and improve the extraction methods and conditions to obtain tannins in an optimal way.

Author Contributions: Conceptualization, M.F.-C., M.A.P. and J.S.-G.; methodology, A.J., J.E., M.C., M.F.-C., P.O. and P.G.-O.; formal analysis, A.J., J.E., M.C., M.F.-C., P.G.-O., P.O. and B.N.-E.; investigation, A.J., J.E., M.C., M.F.-C., P.G.-O., P.O. and B.N.-E.; writing—original draft preparation, A.J., J.E., M.C., M.F.-C., P.G.-O., P.O. and B.N.-E.; writing—review and editing, M.F.-C., M.A.P. and J.S.-G.; supervision, M.F.-C., M.A.P. and J.S.-G.; project administration, M.F.-C., M.A.P. and J.S.-G. All authors have read and agreed to the published version of the manuscript.

Funding: The research leading to these results was funded by FEDER under the program Interreg V-A Spain-Portugal (POPTEC) 2014–2020 ref. 0377_IBERPHENOL_6_E and ref. 0181_NANOEATERS_01_E; to Xunta de Galicia supporting the Axudas Conecta Peme the IN852A 2018/58 NeuroFood Project and the program EXCELENCIA-ED431F 2020/12; to Ibero-American Program on Science and Technology (CYTED—AQUA-CIBUS, P317RT0003) and by the Bio Based Industries Joint Undertaking (JU) under grant agreement No 888003 UP4HEALTH Project (H2020-BBI-JTI-2019). The JU receives support from the European Union's Horizon 2020 research and innovation program and the Bio Based Industries Consortium.

Institutional Review Board Statement: Not applicable.

Informed Consent Statement: Not applicable.

Data Availability Statement: Not applicable.

Acknowledgments: The research leading to these results was supported by MICINN supporting the Ramón y Cajal grant for M.A. Prieto (RYC-2017-22891) and the Juan de la Cierva_incorporación grant for P. Otero (IJCI-2016-27774); by Xunta de Galicia and University of Vigo supporting the post-doctoral grant of M. Fraga-Corral (ED481B-2019/096), the pre-doctoral grant for P. García-Oliveira (ED481A-2019/295); to EcoChestnut Project (Erasmus+ KA202) that supports the work of M. Carpena. The project SYSTEMIC "an integrated approach to the challenge of sustainable food systems: adaptive and mitigatory strategies to address climate change and malnutrition", Knowledge hub on Nutrition and Food Security, has received funding from national research funding parties in Belgium (FWO), France (INRA), Germany (BLE), Italy (MIPAAF), Latvia (IZM), Norway (RCN), Portugal (FCT), and Spain (AEI) in a joint action of JPI HDHL, JPI-OCEANS and FACCE-JPI launched in 2019 under the ERA-NET ERA-HDHL (n° 696295).

Conflicts of Interest: The authors declare no conflict of interest.

Abbreviations

Tannins

CoT	Complex tannin
CT	Condensed tannin
EA	Ellagic acid
ET	Ellagitannin
GA	Gallic acid

GT	Gallotannin
HHDP	Hexahydroxydiphenol
HT	Hydrolysable tannin
PC	Procyanidin
PD	Prodelphinidin
PG	Phloroglucinol
PGG	Pentagalloylglucose
PoGG	Polygalloylglucose
PT	Phlorotannin
TGG	Trigalloylglucose
Bioactivities and Assays	
ADP	Adenosine diphosphate
Bcl-2	Apoptosis inhibitor gen
CAS	Caspase
CAT	Catalase
COX	Cyclooxygenase
CRP	C-reactive protein
DM	Diabetes mellitus
DNA	Deoxyribonucleic acid
GABA	Gamma-aminobutyric acid
GSH	Glutathione (reduced)
HBeAG	E antigen of the hepatitis B virus
HBV	Hepatitis B virus
HIV	Human immunodeficiency virus
ICAM	Intercellular adhesion molecule
IL	Interleukin
iNOS	Nitric oxide synthase
IZ	Inhibition zone
JNK	C-Jun N-terminal kinase
LPS	Lipopolysaccharide
MAD	Malondialdehyde
MBC	Minimum bactericidal concentration
MIC	Minimal inhibitory concentration
MMP	Matrix metalloproteinase
mRNA	Messenger ribonucleic acid
MRSA	Methicillin-resistant *Staphylococcus aureus*
NADPH	Nicotinamide adenine dinucleotide (reduced)
NF-κB	Nuclear factor-κB
NO	Nitric oxide
NSAID	Nonsteroidal anti-inflammatory drug
RNA	Ribonucleic acid
ROS	Reactive oxygen species
SOD	Superoxide dismutase
TNF-α	Tumor necrosis factor-α
VCAM	Vascular cell adhesion protein
VEGF	Vascular endothelial growth factor

References

1. Barbehenn, R.V.; Peter Constabel, C. Tannins in plant-herbivore interactions. *Phytochemistry* **2011**, *72*, 1551–1565. [CrossRef] [PubMed]
2. De Hoyos-Martínez, P.L.; Merle, J.; Labidi, J. Charrier–El Bouhtoury, F. Tannins extraction: A key point for their valorization and cleaner production. *J. Clean. Prod.* **2019**, *206*, 1138–1155. [CrossRef]
3. Shirmohammadli, Y.; Efhamisisi, D.; Pizzi, A. Tannins as a sustainable raw material for green chemistry: A review. *Ind. Crop. Prod.* **2018**, *126*, 316–332. [CrossRef]
4. Vuolo, M.M.; Lima, V.S.; Maróstica Junior, M.R. Phenolic compounds: Structure, classification, and antioxidant power. In *Bioactive Compounds: Health Benefits and Potential Applications*; Elsevier Inc.: Amsterdam, The Netherlands, 2018; pp. 33–50, ISBN 9780128147757.
5. Ky, I.; Le Floch, A.; Zeng, L.; Pechamat, L.; Jourdes, M.; Teissedre, P.L. Tannins. *Encycl. Food Health* **2015**, *7*, 247–255.

6. Okuda, T.; Ito, H. Tannins of constant structure in medicinal and food plants-hydrolyzable tannins and polyphenols related to tannins. *Molecules* **2011**, *16*, 2191–2217. [CrossRef]
7. Macáková, K.; Kolečkář, V.; Cahlíková, L.; Chlebek, J.; Hošťálková, A.; Kuča, K.; Jun, D.; Opletal, L. Tannins and their influence on health. In *Recent Advances in Medicinal Chemistry*; Rahman, A., Choudhary, M., Perry, G., Eds.; Bentham Science Publishers: Sarja, UAE, 2014; Volume 1, pp. 159–208, ISBN 9780128039618.
8. Molino, S.; Casanova, N.A.; Henares, J.Á.R.; Miyakawa, M.E.F. Natural tannin wood extracts as a potential food ingredient in the food industry. *J. Agric. Food Chem.* **2020**, *68*, 2836–2848. [CrossRef]
9. Shahidi, F.; Ambigaipalan, P. Phenolics and polyphenolics in foods, beverages and spices: Antioxidant activity and health effects—A review. *J. Funct. Foods* **2015**, *18*, 820–897. [CrossRef]
10. Fraga-Corral, M.; García-Oliveira, P.; Pereira, A.G.; Lourenço-Lopes, C.; Jimenez-Lopez, C.; Prieto, M.A.; Simal-Gandara, J. Technological application of tannin-based extracts. *Molecules* **2020**, *25*, 614. [CrossRef]
11. Jiménez, N.; Esteban-Torres, M.; Mancheño, J.M.; de las Rivas, B.; Muñoza, R. Tannin degradation by a novel tannase enzyme present in some *Lactobacillus plantarum* strains. *Appl. Environ. Microbiol.* **2014**, *80*, 2991–2997. [CrossRef]
12. Arapitsas, P. Hydrolyzable tannin analysis in food. *Food Chem.* **2012**, *135*, 1708–1717. [CrossRef]
13. Hagerman, A.E. Hydrolyzable tannin structural chemistry. *Tann. Handb.* **2010**, 1–8. Available online: http//www.users.muohio.edu/hagermae/tannin.pdf (accessed on 11 October 2020).
14. Rousserie, P.; Rabot, A.; Geny-Denis, L. From flavanols biosynthesis to wine tannins: What place for grape seeds? *J. Agric. Food Chem.* **2019**, *67*, 1325–1343. [CrossRef] [PubMed]
15. Sieniawska, E.; Baj, T. Tannins. In *Pharmacognosy*; Elsevier: Amsterdam, The Netherlands, 2017; pp. 199–232.
16. Venkatesan, J.; Keekan, K.K.; Anil, S.; Bhatnagar, I.; Kim, S.-K. Phlorotannins. *Encycl. Food Chem.* **2019**, 515–527. [CrossRef]
17. Erpel, F.; Mateos, R.; Pérez-Jiménez, J.; Pérez-Correa, J.R. Phlorotannins: From isolation and structural characterization, to the evaluation of their antidiabetic and anticancer potential. *Food Res. Int.* **2020**, *137*, 109589. [CrossRef]
18. Salminen, J.P.; Karonen, M. Chemical ecology of tannins and other phenolics: We need a change in approach. *Funct. Ecol.* **2011**, *25*, 325–338. [CrossRef]
19. Huang, Q.; Liu, X.; Zhao, G.; Hu, T.; Wang, Y. Potential and challenges of tannins as an alternative to in-feed antibiotics for farm animal production. *Anim. Nutr.* **2018**, *4*, 137–150. [CrossRef]
20. Sharma, K.; Kumar, V.; Kaur, J.; Tanwar, B.; Goyal, A.; Sharma, R.; Gat, Y.; Kumar, A. Health effects, sources, utilization and safety of tannins: A critical review. *Toxin Rev.* **2019**, 1–13. [CrossRef]
21. Sharma, K.; Guleria, S.; Razdan, V.K.; Babu, V. Synergistic antioxidant and antimicrobial activities of essential oils of some selected medicinal plants in combination and with synthetic compounds. *Ind. Crop. Prod.* **2020**, *154*, 112569. [CrossRef]
22. Rana, J.; Paul, J. Consumer behavior and purchase intention for organic food: A review and research agenda. *J. Retail. Consum. Serv.* **2017**, *38*, 157–165. [CrossRef]
23. Winans, K.; Kendall, A.; Deng, H. The history and current applications of the circular economy concept. *Renew. Sustain. Energy Rev.* **2017**, *68*, 825–833. [CrossRef]
24. Jimenez-Lopez, C.; Fraga-Corral, M.; Carpena, M.; García-Oliveira, P.; Echave, J.; Pereira, A.G.; Lourenço-Lopes, C.; Prieto, M.A.; Simal-Gandara, J. Agriculture waste valorisation as a source of antioxidant phenolic compounds within a circular and sustainable bioeconomy. *Food Funct.* **2020**, *11*, 4853–4877. [CrossRef] [PubMed]
25. Aires, A.; Carvalho, R.; Saavedra, M.J. Valorization of solid wastes from chestnut industry processing: Extraction and optimization of polyphenols, tannins and ellagitannins and its potential for adhesives, cosmetic and pharmaceutical industry. *Waste Manag.* **2016**, *48*, 457–464. [CrossRef] [PubMed]
26. Grenda, K.; Arnold, J.; Hunkeler, D.; Gamelas, J.A.F.; Rasteiro, M.G. Tannin-based coagulants from laboratory to pilot plant scales for coloured wastewater treatment. *BioResources* **2018**, *13*, 2727–2747. [CrossRef]
27. Cassani, L.; Gomez-zavaglia, A.; Jimenez-lopez, C.; Prieto, M.A.; Simal-gandara, J. Seaweed-based natural ingredients: Stability of phlorotannins during extraction, storage, passage through the gastrointestinal tract and potential incorporation into functional foods. *Food Res. Int.* **2020**, 109676. [CrossRef] [PubMed]
28. Das, A.K.; Islam, M.N.; Faruk, M.O.; Ashaduzzaman, M.; Dungani, R. Review on tannins: Extraction processes, applications and possibilities. *S. Afr. J. Bot.* **2020**, *135*, 58–70. [CrossRef]
29. Mirabella, N.; Castellani, V.; Sala, S. Current options for the valorization of food manufacturing waste: A review. *J. Clean. Prod.* **2014**, *65*, 28–41. [CrossRef]
30. Teixeira, A.; Baenas, N.; Dominguez-Perles, R.; Barros, A.; Rosa, E.; Moreno, D.A.; Garcia-Viguera, C. Natural bioactive compounds from winery by-products as health promoters: A review. *Int. J. Mol. Sci.* **2014**, *15*, 15638–15678. [CrossRef]
31. Smeriglio, A.; Barreca, D.; Bellocco, E.; Trombetta, D. Proanthocyanidins and hydrolysable tannins: Occurrence, dietary intake and pharmacological effects. *Br. J. Pharmacol.* **2017**, *174*, 1244–1262. [CrossRef]
32. Serrano, J.; Puupponen-Pimiä, R.; Dauer, A.; Aura, A.-M.; Saura-Calixto, F. Tannins: Current knowledge of food sources, intake, bioavailability and biological effects. *Mol. Nutr. Food Res.* **2009**, *53*, S310–S329. [CrossRef]
33. Xiong, J.; Grace, M.H.; Esposito, D.; Wang, F.; Lila, M.A. Phytochemical characterization and anti-inflammatory properties of *Acacia mearnsii* leaves. *Nat. Prod. Commun.* **2016**, *11*. [CrossRef]

dos Santos, C.; Vargas, Á.; Fronza, N.; dos Santos, J.H.Z. Structural, textural and morphological characteristics of tannins from *Acacia mearnsii* encapsulated using sol-gel methods: Applications as antimicrobial agents. *Colloids Surf. B Biointerfaces* **2017**, *151*, 26–33. [CrossRef]

Rather, L.J.; Shahid-ul-Islam; Mohammad, F. *Acacia nilotica* (L.): A review of its traditional uses, phytochemistry, and pharmacology. *Sustain. Chem. Pharm.* **2015**, *2*, 12–30. [CrossRef]

Safari, V.Z.; Kamau, J.K.; Nthiga, P.M.; Ngugi, M.P.; Orinda, G.; Njagi, E.M. Antipyretic, antiinflammatory and antinociceptive activities of aqueous bark extract of *Acacia nilotica* (L.) Delile in albino mice. *Pain Manag. Med.* **2016**, *2*, 2.

Brizi, C.; Santulli, C.; Micucci, M.; Budriesi, R.; Chiarini, A.; Aldinucci, C.; Frosini, M. Neuroprotective Effects of *Castanea sativa* Mill. bark extract in human neuroblastoma cells subjected to oxidative stress. *J. Cell. Biochem.* **2016**, *117*, 510–520. [CrossRef] [PubMed]

Chiocchio, I.; Prata, C.; Mandrone, M.; Ricciardiello, F.; Marrazzo, P.; Tomasi, P.; Angeloni, C.; Fiorentini, D.; Malaguti, M.; Poli, F.; et al. Leaves and spiny burs of *Castanea Sativa* from an experimental chestnut grove: Metabolomic analysis and anti-neuroinflammatory activity. *Metabolites* **2020**, *10*, 408. [CrossRef]

Cardullo, N.; Muccilli, V.; Saletti, R.; Giovando, S.; Tringali, C. A mass spectrometry and 1H NMR study of hypoglycemic and antioxidant principles from a *Castanea sativa* tannin employed in oenology. *Food Chem.* **2018**, *268*, 585–593. [CrossRef] [PubMed]

Chiarini, A.; Micucci, M.; Malaguti, M.; Budriesi, R.; Ioan, P.; Lenzi, M.; Fimognari, C.; Gallina Toschi, T.; Comandini, P.; Hrelia, S. Sweet chestnut (*Castanea sativa* Mill.) bark extract: Cardiovascular activity and myocyte protection against oxidative damage. *Oxidative Med. Cell. Longev.* **2013**, *2013*, 471790. [CrossRef]

Silva, V.; Falco, V.; Dias, M.I.; Barros, L.; Silva, A.; Capita, R.; Alonso-Calleja, C.; Amaral, J.S.; Igrejas, G.; Ferreira, I.C.F.R.; et al. Evaluation of the phenolic profile of *Castanea sativa* Mill. by-products and their antioxidant and antimicrobial activity against multiresistant bacteria. *Antioxidants* **2020**, *9*, 87. [CrossRef]

Štumpf, S.; Hostnik, G.; Primožič, M.; Leitgeb, M.; Salminen, J.-P.; Bren, U. The effect of growth medium strength on minimum inhibitory concentrations of tannins and tannin extracts against *E. coli*. *Molecules* **2020**, *25*, 2947. [CrossRef]

Reggi, S.; Giromini, C.; Dell'Anno, M.; Baldi, A.; Rebucci, R.; Rossi, L. In vitro digestion of chestnut and quebracho tannin extracts: Antimicrobial effect, antioxidant capacity and cytomodulatory activity in swine intestinal IPEC-J2 cells. *Animals* **2020**, *10*, 195. [CrossRef]

Esposito, T.; Celano, R.; Pane, C.; Piccinelli, A.L.; Sansone, F.; Picerno, P.; Zaccardelli, M.; Aquino, R.P.; Mencherini, T. Chestnut (*Castanea sativa* miller.) burs extracts and functional compounds: Uhplc-uv-hrms profiling, antioxidant activity, and inhibitory effects on phytopathogenic fungi. *Molecules* **2019**, *24*, 302. [CrossRef] [PubMed]

Bonelli, F.; Turini, L.; Sarri, G.; Serra, A.; Buccioni, A.; Mele, M. Oral administration of chestnut tannins to reduce the duration of neonatal calf diarrhea. *BMC Vet. Res.* **2018**, *14*, 4–9. [CrossRef] [PubMed]

Regueiro, J.; Sánchez-González, C.; Vallverdú-Queralt, A.; Simal-Gándara, J.; Lamuela-Raventós, R.; Izquierdo-Pulido, M. Comprehensive identification of walnut polyphenols by liquid chromatography coupled to linear ion trap-Orbitrap mass spectrometry. *Food Chem.* **2014**, *152*, 340–348. [CrossRef] [PubMed]

Papoutsi, Z.; Kassi, E.; Chinou, I.; Halabalaki, M.; Skaltsounis, L.A.; Moutsatsou, P. Walnut extract (*Juglans regia* L.) and its component ellagic acid exhibit anti-inflammatory activity in human aorta endothelial cells and osteoblastic activity in the cell line KS483. *Br. J. Nutr.* **2008**, *99*, 715–722. [CrossRef]

Sun, Y.; Qi, G.; Li, D.; Meng, H.; Zhu, Z.; Zhao, Y.; Qi, Y.; Zhang, X. Walnut (*Juglans regia* L.) kernel extracts protect against isoproterenol-induced myocardial infarction in rats. *Rejuvenation Res.* **2018**, *22*, 306–312. [CrossRef]

Meagher, L.P.; Lane, G.; Sivakumaran, S.; Tavendale, M.H.; Fraser, K. Characterization of condensed tannins from *Lotus* species by thiolytic degradation and electrospray mass spectrometry. *Anim. Feed Sci. Technol.* **2004**, *117*, 151–163. [CrossRef]

Hedqvist, H.; Mueller-Harvey, I.; Reed, J.D.; Krueger, C.G.; Murphy, M. Characterisation of tannins and in vitro protein digestibility of several *Lotus corniculatus* varieties. *Anim. Feed Sci. Technol.* **2000**, *87*, 41–56. [CrossRef]

Min, B.R.; Fernandez, J.M.; Barry, T.N.; McNabb, W.C.; Kemp, P.D. The effect of condensed tannins in Lotus corniculatus upon reproductive efficiency and wool production in ewes during autumn. *Anim. Feed Sci. Technol.* **2001**, *92*, 185–202. [CrossRef]

Raitanen, J.-E.; Järvenpää, E.; Korpinen, R.; Mäkinen, S.; Hellström, J.; Kilpeläinen, P.; Liimatainen, J.; Ora, A.; Tupasela, T.; Jyske, T. Tannins of conifer bark as nordic piquancy—Sustainable preservative and aroma? *Molecules* **2020**, *25*, 567. [CrossRef]

Liu, C.; Cai, D.; Zhang, L.; Tang, W.; Yan, R.; Guo, H.; Chen, X. Identification of hydrolyzable tannins (punicalagin, punicalin and geraniin) as novel inhibitors of hepatitis B virus covalently closed circular DNA. *Antiviral Res.* **2016**, *134*, 97–107. [CrossRef]

Arunkumar, J.; Rajarajan, S. Study on antiviral activities, drug-likeness and molecular docking of bioactive compounds of *Punica granatum* L. to Herpes simplex virus-2 (HSV-2). *Microb. Pathog.* **2018**, *118*, 301–309. [CrossRef] [PubMed]

Muccilli, V.; Cardullo, N.; Spatafora, C.; Cunsolo, V.; Tringali, C. α-Glucosidase inhibition and antioxidant activity of an oenological commercial tannin. Extraction, fractionation and analysis by HPLC/ESI-MS/MS and 1H NMR. *Food Chem.* **2017**, *215*, 50–60. [CrossRef] [PubMed]

Horvathova, M.; Orszaghova, Z.; Laubertova, L.; Vavakova, M.; Sabaka, P.; Rohdewald, P.; Durackova, Z.; Muchova, J. Effect of the French oak wood extract robuvit on markers of oxidative stress and activity of antioxidant enzymes in healthy volunteers: A pilot study. *Oxidative Med. Cell. Longev.* **2014**, *2014*. [CrossRef] [PubMed]

57. Natella, F.; Leoni, G.; Maldini, M.; Natarelli, L.; Comitato, R.; Schonlau, F.; Virgili, F.; Canali, R. Absorption, metabolis and effects at transcriptome level of a standardized french oak wood extract, Robuvit, in healthy volunteers: Pilot study. *J. Agr Food Chem.* **2014**, *62*, 443–453. [CrossRef] [PubMed]
58. Khalilpour, S.; Sangiovanni, E.; Piazza, S.; Fumagalli, M.; Beretta, G.; Dell'Agli, M. In vitro evidences of the traditional use *Rhus coriaria* L. fruits against skin inflammatory conditions. *J. Ethnopharmacol.* **2019**, *238*, 111829. [CrossRef]
59. Isik, S.; Tayman, C.; Cakir, U.; Koyuncu, I.; Taskin Turkmenoglu, T.; Cakir, E. Sumac (*Rhus coriaria*) for the prevention a treatment of necrotizing enterocolitis. *J. Food Biochem.* **2019**, *43*, e13068. [CrossRef]
60. Monforte, M.T.; Smeriglio, A.; Germanò, M.P.; Pergolizzi, S.; Circosta, C.; Galati, E.M. Evaluation of antioxidant, antiinflammato and gastroprotective properties of *Rubus fruticosus* L. fruit juice. *Phytother. Res.* **2018**, *32*, 1404–1414. [CrossRef]
61. Mirazi, N.; Hosseini, A. Attenuating properties of *Rubus fruticosus* L. on oxidative damage and inflammatory response followi streptozotocin-induced diabetes in the male Wistar rats. *J. Diabetes Metab. Disord.* **2020**. [CrossRef]
62. Li, Y.; Fu, X.; Duan, D.; Liu, X.; Xu, J.; Gao, X. Extraction and identification of phlorotannins from the brown alga, *Sarg sum fusiforme* (Harvey) Setchell. *Mar. Drugs* **2017**, *15*, 49. [CrossRef]
63. Casas, M.P.; Rodríguez-Hermida, V.; Pérez-Larrán, P.; Conde, E.; Liveri, M.T.; Ribeiro, D.; Fernandes, E.; Domínguez, H. In vi bioactive properties of phlorotannins recovered from hydrothermal treatment of *Sargassum muticum*. *Sep. Purif. Technol.* **2016**, *1* 117–126. [CrossRef]
64. Venter, P.B.; Sisa, M.; Van Der Merwe, M.J.; Bonnet, S.L.; van der Westhuizen, J.H. Analysis of commercial proanthocyanidi Part 1: The chemical composition of quebracho (*Schinopsis lorentzii* and *Schinopsis balansae*) heartwood extract. *Phytochemist* **2012**, *73*, 95–105. [CrossRef] [PubMed]
65. Cardullo, N.; Muccilli, V.; Cunsolo, V.; Tringali, C. Mass spectrometry and 1H-NMR study of *Schinopsis lorentzii* (Quebrach tannins as a source of hypoglycemic and antioxidant principles. *Molecules* **2020**, *25*, 3257. [CrossRef] [PubMed]
66. Fruet, A.P.B.; Giotto, F.M.; Fonseca, M.A.; Nörnberg, J.L.; de Mello, A.S. Effects of the incorporation of tannin extract fro quebracho colorado wood on color parameters, lipid oxidation, and sensory attributes of beef patties. *Foods* **2020**, *9*, 6 [CrossRef] [PubMed]
67. Falcão, L.; Araújo, M.E.M. Vegetable tannins used in the manufacture of historic leathers. *Molecules* **2018**, *23*, 1081. [CrossRef] [PubMe
68. Yang, M.H.; Ali, Z.; Khan, I.A.; Khan, S.I. Anti-inflammatory activity of constituents isolated from *Terminalia chebu Nat. Prod. Commun.* **2014**, *9*. [CrossRef]
69. Rodríguez-Pérez, C.; García-Villanova, B.; Guerra-Hernández, E.; Verardo, V. Grape seeds proanthocyanidins: An overview in vivo bioactivity in animal models. *Nutrients* **2019**, *11*, 2435. [CrossRef]
70. Sieniawska, E. Activities of Tannins—From in vitro studies to clinical trials. *Nat. Prod. Commun.* **2015**, *10*. [CrossRef]
71. Daré, R.G.; Nakamura, C.V.; Ximenes, V.F.; Lautenschlager, S.O.S. Tannic acid, a promising anti-photoaging agent: Evidences its antioxidant and anti-wrinkle potentials, and its ability to prevent photodamage and MMP-1 expression in L929 fibrobla exposed to UVB. *Free Radic. Biol. Med.* **2020**, *160*, 342–355. [CrossRef]
72. Okuda, T.; Kimura, Y.; Yoshida, T.; Hatano, T.; Okuda, H.; Arichi, S. Studies on the activities of tannins and related co pounds from medicinal plants and drugs. I. Inhibitory effects on lipid peroxidation in mitochondria and microsomes of liv *Chem. Pharm. Bull.* **1983**, *31*, 1625–1631. [CrossRef]
73. Iwata, S.; Fukaya, Y.; Nakazawa, K.; Okuda, T. Effects of tannins on the oxidative damage of mouse ocular lens I. using t oxidative damage model induced by the xanthine-xanthine oxidase system. *J. Ocul. Pharmacol. Ther.* **1987**, *3*, 227–238. [CrossRe
74. Beninger, C.W.; Hosfield, G.L. Antioxidant activity of extracts, condensed tannin fractions, and pure flavonoids from *Phase lus vulgaris* L. seed coat color genotypes. *J. Agric. Food Chem.* **2003**, *51*, 7879–7883. [CrossRef] [PubMed]
75. Tian, Y.; Zou, B.; Li, C.; Yang, J.; Xu, S.; Hagerman, A.E. High molecular weight persimmon tannin is a potent antioxidant both vivo and in vivo. *Food Res. Int.* **2012**, *45*, 26–30. [CrossRef]
76. Huang, W.; Niu, H.; Xue, X.; Li, J.; Li, C. Robinetinidol-(4β→8)-epigallocatechin 3-O-gallate, a galloyl dimer prorobinetinidin fro *Acacia mearnsii* De Wild, effectively protects human neuroblastoma SH-SY5Y cells against acrolein-induced oxidative damag *J. Alzheimer's Dis.* **2010**, *21*, 493–506. [CrossRef] [PubMed]
77. Terra, X.; Montagut, G.; Bustos, M.; Llopiz, N.; Ardèvol, A.; Bladé, C.; Fernández-Larrea, J.; Pujadas, G.; Salvadó, J.; Arola, L.; et Grape-seed procyanidins prevent low-grade inflammation by modulating cytokine expression in rats fed a high-fat di *J. Nutr. Biochem.* **2009**, *20*, 210–218. [CrossRef] [PubMed]
78. Park, M.; Cho, H.; Jung, H.; Lee, H.; Hwang, K.T. Antioxidant and anti-inflammatory activities of tannin fraction of the extra from black raspberry seeds compared to grape seeds. *J. Food Biochem.* **2014**. [CrossRef]
79. Kowalska, K.; Olejnik, A.; Zielińska-Wasielica, J.; Olkowicz, M. Raspberry (*Rubus idaeus* L.) fruit extract decreases oxidati markers, improves lipid metabolism and reduces adipose tissue inflammation in hypertrophied 3T3-L1 adipocytes. *J. Funct. Foo* **2019**, *62*, 103568. [CrossRef]
80. Li, D.Q.; Zhao, J.; Xie, J.; Li, S.P. A novel sample preparation and on-line HPLC-DAD-MS/MS-BCD analysis for rapid screeni and characterization of specific enzyme inhibitors in herbal extracts: Case study of α-glucosidase. *J. Pharm. Biomed. Anal.* **201** *88*, 130–135. [CrossRef]
81. Senthilkumar, G.P.; Subramanian, S.P. Biochemical studies on the effect of *Terminalia chebula* on the levels of glycoproteins streptozotocin-induced experimental diabetes in rats. *J. Appl. Biomed.* **2008**, *6*, 105–115. [CrossRef]

82. Matsumoto, K.; Yokoyama, S. ichiro Induction of uncoupling protein-1 and -3 in brown adipose tissue by kaki-tannin in type 2 diabetic NSY/Hos mice. *Food Chem. Toxicol.* **2012**, *50*, 184–190. [CrossRef]
83. Dudzinska, D.; Bednarska, K.; Boncler, M.; Luzak, B.; Watala, C. The influence of *Rubus idaeus* and *Rubus caesius* leaf extracts on platelet aggregation in whole blood. Cross-talk of platelets and neutrophils. *Platelets* **2016**, *27*, 433–439. [CrossRef]
84. Ram, H.; Jatwa, R.; Purohit, A. Antiatherosclerotic and cardioprotective potential of *Acacia senegal* seeds in diet-induced atherosclerosis in rabbits. *Biochem. Res. Int.* **2014**, *2014*, 436848. [CrossRef] [PubMed]
85. Tonin, T.D.; Thiesen, L.C.; de Oliveira Nunes, M.L.; Broering, M.F.; Donato, M.P.; Goss, M.J.; Petreanu, M.; Niero, R.; Machado, I.D.; Santin, J.R. *Rubus imperialis* (Rosaceae) extract and pure compound niga-ichigoside F1: Wound healing and anti-inflammatory effects. *Naunyn. Schmiedeberg's. Arch. Pharmacol.* **2016**, *389*, 1235–1244. [CrossRef] [PubMed]
86. Gabr, S.A.; Alghadir, A.H. Evaluation of the biological effects of lyophilized hydrophilic extract of *rhus coriaria* on Myeloperoxidase (MPO) activity, wound healing, and microbial infections of skin wound tissues. *Evid. Based Complement. Altern. Med.* **2019**, *2019*, 5861537. [CrossRef] [PubMed]
87. Quave, C.L.; Plano, L.R.W.; Pantuso, T.; Bennett, B.C. Effects of extracts from Italian medicinal plants on planktonic growth, biofilm formation and adherence of methicillin-resistant *Staphylococcus aureus*. *J. Ethnopharmacol.* **2008**, *118*, 418–428. [CrossRef] [PubMed]
88. Saraiva, A.M.; Saraiva, C.L.; Cordeiro, R.P.; Soares, R.R.; Xavier, H.S.; Caetano, N. Atividade antimicrobiana e sinérgica das frações das folhas de *Schinopsis brasiliensis* Engl. frente a clones multirresistentes de *Staphylococcus aureus*. *Rev. Bras. Plantas Med.* **2013**, *15*, 199–207. [CrossRef]
89. Liu, M.; Feng, M.; Yang, K.; Cao, Y.; Zhang, J.; Xu, J.; Hernández, S.H.; Wei, X.; Fan, M. Transcriptomic and metabolomic analyses reveal antibacterial mechanism of astringent persimmon tannin against Methicillin-resistant *Staphylococcus aureus* isolated from pork. *Food Chem.* **2020**, *309*, 125692. [CrossRef]
90. Martini, S.; D'Addario, C.; Colacevich, A.; Focardi, S.; Borghini, F.; Santucci, A.; Figura, N.; Rossi, C. Antimicrobial activity against *Helicobacter pylori* strains and antioxidant properties of blackberry leaves (*Rubus ulmifolius*) and isolated compounds. *Int. J. Antimicrob. Agents* **2009**, *34*, 50–59. [CrossRef]
91. Hajaji, S.; Jabri, M.-A.; Sifaoui, I.; López-Arencibia, A.; Reyes-Batlle, M.; B'chir, F.; Valladares, B.; Pinero, J.E.; Lorenzo-Morales, J.; Akkari, H. Amoebicidal, antimicrobial and in vitro ROS scavenging activities of Tunisian *Rubus ulmifolius* Schott, methanolic extract. *Exp. Parasitol.* **2017**, *183*, 224–230. [CrossRef]
92. Sisti, M.; De Santi, M.; Fraternale, D.; Ninfali, P.; Scoccianti, V.; Brandi, G. Antifungal activity of *Rubus ulmifolius* Schott standardized *in vitro* culture. *LWT Food Sci. Technol.* **2008**, *41*, 946–950. [CrossRef]
93. Elizondo, A.M.; Mercado, E.C.; Rabinovitz, B.C.; Fernandez-Miyakawa, M.E. Effect of tannins on the in vitro growth of *Clostridium perfringens*. *Vet. Microbiol.* **2010**, *145*, 308–314. [CrossRef]
94. Khalifa, I.; Zhu, W.; Mohammed, H.H.H.; Dutta, K.; Li, C. Tannins inhibit SARS-CoV-2 through binding with catalytic dyad residues of 3CLpro: An in silico approach with 19 structural different hydrolysable tannins. *J. Food Biochem.* **2020**, *44*, e13432. [CrossRef] [PubMed]
95. Subbotina, M.D.; Timchenko, V.N.; Vorobyov, M.M.; Konunova, Y.S.; Aleksandrovih, Y.S.; Shushunov, S. Effect of oral administration of tormentil root extract (Potentilla tormentilla) on rotavirus diarrhea in children: A randomized, double blind, controlled trial. *Pediatr. Infect. Dis. J.* **2003**, *22*, 706–711. [CrossRef] [PubMed]
96. Russo, M.; Coppola, V.; Giannetti, E.; Buonavolontà, R.; Piscitelli, A.; Staiano, A. Oral administration of tannins and flavonoids in children with acute diarrhea: A pilot, randomized, control-case study. *Ital. J. Pediatrics* **2018**, *44*, 4–9. [CrossRef] [PubMed]
97. Molino, S.; Fernández-Miyakawa, M.; Giovando, S.; Rufián-Henares, J.Á. Study of antioxidant capacity and metabolization of quebracho and chestnut tannins through in vitro gastrointestinal digestion-fermentation. *J. Funct. Foods* **2018**, *49*, 188–195. [CrossRef]
98. Noce, A.; Marrone, G.; Bernini, R.; Campo, M.; Di Lauro, M.; Pietroboni Zaitseva, A.; Romani, A.; di Daniele, N. Impact of tannins as food supplement in a ckd population with recurrent urinary tract infections: Preliminary data. *Nephrol. Dial. Transplant.* **2020**, *35*, 1264. [CrossRef]
99. Chandrasekhar, Y.; Phani Kumar, G.; Navya, K.; Ramya, E.M.; Anilakumar, K.R. Tannins from Terminalia chebula fruits attenuates GABA antagonist-induced anxiety-like behaviour via modulation of neurotransmitters. *J. Pharm. Pharmacol.* **2018**, *70*, 1662–1674. [CrossRef]
100. Fagundes, G.M.; Benetel, G.; Santos, K.C.; Welter, K.C.; Melo, F.A.; Muir, J.P.; Bueno, I.C.S. Tannin-rich plants as natural manipulators of rumen fermentation in the livestock industry. *Molecules* **2020**, *25*, 2943. [CrossRef]
101. El-Saber Batiha, G.; Beshbishy, A.M.; Adeyemi, O.S.; Nadwa, E.H.; Rashwan, E.K.M.; Alkazmi, L.M.; Elkelish, A.A.; Igarashi, I. Phytochemical screening and antiprotozoal effects of the methanolic berberis vulgaris and acetonic rhus coriaria extracts. *Molecules* **2020**, *25*, 550. [CrossRef]
102. Quattrucci, A.; Ovidi, E.; Tiezzi, A.; Vinciguerra, V.; Balestra, G.M. Biological control of tomato bacterial speck using Punica granatum fruit peel extract. *Crop Prot.* **2013**, *46*, 18–22. [CrossRef]
103. Li, R.; Dai, T.; Zhou, W.; Fu, G.; Wan, Y.; McClements, D.J.; Li, J. Impact of pH, ferrous ions, and tannic acid on lipid oxidation in plant-based emulsions containing saponin-coated flaxseed oil droplets. *Food Res. Int.* **2020**, *136*, 109618. [CrossRef]
104. Al-Hijazeen, M.; Lee, E.J.; Mendonca, A.; Ahn, D.U. Effects of tannic acid on lipid and protein oxidation, color, and volatiles of raw and cooked chicken breast meat during storage. *Antioxidants* **2016**, *5*, 19. [CrossRef] [PubMed]

105. Engelberth, A.S. Evaluating economic potential of food waste valorization: Onward to a diverse feedstock biorefinery. *Curr. Opin Green Sustain. Chem.* **2020**, *26*, 100385. [CrossRef]
106. Makkar, H.P.S.; Singh, B.; Negi, S.S. Tannin levels and their degree of polymerisation and specific activity in some agro-industria by-products. *Biol. Wastes* **1990**, *31*, 137–144. [CrossRef]
107. Si Heung Sung Antibacterial and antioxidant activities of tannins extracted from agricultural by-products. *J. Med. Plants Res* **2012**, *6*, 3072–3079. [CrossRef]
108. Tavares, C.S.; Martins, A.; Miguel, M.G.; Carvalheiro, F.; Duarte, L.C.; Gameiro, J.A.; Figueiredo, A.C.; Roseiro, L.B. Bioprod ucts from forest biomass II. Bioactive compounds from the steam-distillation by-products of *Cupressus lusitanica* Mill. an *Cistus ladanifer* L. wastes. *Ind. Crop. Prod.* **2020**, *158*, 112991. [CrossRef]
109. Giovando, S.; Koch, G.; Romagnoli, M.; Paul, D.; Vinciguerra, V.; Tamantini, S.; Marini, F.; Zikeli, F.; Mugnozza, G.S. Spectro topochemical investigation of the location of polyphenolic extractives (tannins) in chestnut wood structure and ultrastructure *Ind. Crop. Prod.* **2019**, *141*, 111767. [CrossRef]
110. Bhoite, R.N.; Murthy, P.S. Biodegradation of coffee pulp tannin by *Penicillium verrucosum* for production of tannase, statistica optimization and its application. *Food Bioprod. Process.* **2015**, *94*, 727–735. [CrossRef]
111. Arbenz, A.; Avérous, L. Chemical modification of tannins to elaborate aromatic biobased macromolecular architectures *Green Chem.* **2015**, *17*, 2626–2646. [CrossRef]
112. Liang, X.; Jiang, Y.; Guo, Z.; Fang, S. Separation, UPLC-QTOF-MS/MS analysis, and antioxidant activity of hydrolyzable tannin from water caltrop (*Trapa quadrispinosa*) pericarps. *LWT* **2020**, *133*, 110010. [CrossRef]
113. Wong-Paz, J.E.; Guyot, S.; Aguilar-Zárate, P.; Muñiz-Márquez, D.B.; Contreras-Esquivel, J.C.; Aguilar, C.N. Structural characteri zation of native and oxidized procyanidins (condensed tannins) from coffee pulp (*Coffea arabica*) using phloroglucinolysis and thioglycolysis-HPLC-ESI-MS. *Food Chem.* **2021**, *340*, 127830. [CrossRef]
114. Hayder, Z.; Elfalleh, W.; Othman, K.; Benabderrahim, M.A.; Hannachi, H. Modeling of polyphenols extraction from pomegranate by-product using rotatable central composite design of experiments. *Acta Ecol. Sin.* **2020**. In Press. [CrossRef]
115. Ping, L.; Pizzi, A.; Guo, Z.D.; Brosse, N. Condensed tannins from grape pomace: Characterization by FTIR and MALDI TOF and production of environment friendly wood adhesive. *Ind. Crop. Prod.* **2012**, *40*, 13–20. [CrossRef]
116. Hashim, N.; Paramasivam, M.; Tan, J.S.; Kernain, D.; Hussin, M.H.; Brosse, N.; Gambier, F.; Raja, P.B. Green mode synthesis o silver nanoparticles using *Vitis vinifera*'s tannin and screening its antimicrobial activity / apoptotic potential versus cancer cells *Mater. Today Commun.* **2020**, *25*, 101511. [CrossRef]
117. Cerulli, A.; Napolitano, A.; Masullo, M.; Hošek, J.; Pizza, C.; Piacente, S. Chestnut shells (Italian cultivar "Marrone di Roc cadaspide" PGI): Antioxidant activity and chemical investigation with in depth LC-HRMS/MSn rationalization of tannins *Food Res. Int.* **2020**, *129*, 108787. [CrossRef]
118. Saad, H.; Charrier-El Bouhtoury, F.; Pizzi, A.; Rode, K.; Charrier, B.; Ayed, N. Characterization of pomegranate peels tannin extractives. *Ind. Crop. Prod.* **2012**, *40*, 239–246. [CrossRef]
119. Maran, J.P.; Manikandan, S.; Priya, B.; Gurumoorthi, P. Box-Behnken design based multi-response analysis and optimization of supercritical carbon dioxide extraction of bioactive flavonoid compounds from tea (*Camellia sinensis* L.) leaves. *J. Food Sci. Technol* **2015**, *52*, 92–104. [CrossRef]
120. Naima, R.; Oumam, M.; Hannache, H.; Sesbou, A.; Charrier, B.; Pizzi, A.; El Bouhtoury, F.C. Comparison of the impact of different extraction methods on polyphenols yields and tannins extracted from Moroccan *Acacia mollissima* barks. *Ind. Crop. Prod.* **2015**, *70* 245–252. [CrossRef]
121. Dahmoune, F.; Nayak, B.; Moussi, K.; Remini, H.; Madani, K. Optimization of microwave-assisted extraction of polyphenols from *Myrtus communis* L. leaves. *Food Chem.* **2015**, *166*, 585–595. [CrossRef]
122. Politi, F.A.S.; de Mello, J.C.P.; Migliato, K.F.; Nepomuceno, A.L.A.; Moreira, R.R.D.; Pietro, R.C.L.R. Antimicrobial, cytotoxic and antioxidant activities and determination of the total tannin content of bark extracts *Endopleura uchi*. *Int. J. Mol. Sci.* **2011**, *12* 2757–2768. [CrossRef]
123. Kemppainen, K.; Siika-aho, M.; Pattathil, S.; Giovando, S.; Kruus, K. Spruce bark as an industrial source of condensed tannins and non-cellulosic sugars. *Ind. Crop. Prod.* **2014**, *52*, 158–168. [CrossRef]
124. Talmaciu, A.I.; Ravber, M.; Volf, I.; Knez, Ž.; Popa, V.I. Isolation of bioactive compounds from spruce bark waste using sub- and supercritical fluids. *J. Supercrit. Fluids* **2016**, *117*, 243–251. [CrossRef]
125. Liu, Z.; Chen, Z.; Han, F.; Kang, X.; Gu, H.; Yang, L. Microwave-assisted method for simultaneous hydrolysis and extrac tion in obtaining ellagic acid, gallic acid and essential oil from *Eucalyptus globulus* leaves using Brönsted acidic ionic liquid [HO3S(CH2)4mim]HSO4. *Ind. Crop. Prod.* **2016**, *81*, 152–161. [CrossRef]

foods MDPI

Review

Rice Bran By-Product: From Valorization Strategies to Nutritional Perspectives

Marco Spaggiari [1], Chiara Dall'Asta [1], Gianni Galaverna [1] and María Dolores del Castillo Bilbao [2,*]

[1] Department of Food and Drug, University of Parma, Parco Area delle Scienze, 17/A, 43121 Parma, Italy; marco.spaggiari1@studenti.unipr.it (M.S.); chiara.dallasta@unipr.it (C.D.); gianni.galaverna@unipr.it (G.G.)
[2] Food Bioscience Group, Department of Bioactivity and Food Analysis, Instituto de Investigación en Ciencias de la Alimentación (CIAL) (CSIC-UAM), Calle Nicolás Cabrera, 9, 28049 Madrid, Spain
* Correspondence: mdolores.delcastillo@csic.es; Tel.: +34-91-0017900 (ext. 953)

Abstract: The aim of this study is to review the innovative techniques based on bioprocessing, thermal or physical treatments which have been proposed during the last few decades to convert rice bran into a valuable food ingredient. Rice bran (*Oryza sativa*) is the main by-product of rice grain processing. It is produced in large quantities worldwide and it contains a high amount of valuable nutrients and bioactive compounds with significant health-related properties. Despite that, its application in food industry is still scarce because of its sensitivity to oxidation processes, instability and poor technological suitability. Furthermore, the health-related effects of pretreated rice bran are also presented in this review, considering the up-to-date literature focused on both in vivo and in vitro studies. Moreover, in relation to this aspect, a brief description of rice bran arabinoxylans is provided. Finally, the application of rice bran in the food industry and the main technology aspects are concisely summarized.

Keywords: rice bran; by-products; bioprocessing; bioactive compounds; nutritional value; rice bran arabinoxylans

1. Introduction

Food processing is a set of operations that permit to transform raw materials into valuable food ingredients. Cereal crops, for example, are rarely consumed as whole grains and during their transformation process a large amount of residue is produced. Rice bran (RB) is a by-product that derives from the milling of rice grain, the third most consumed cereal overworld [1]. It represents around 12% of the total kernel weight and it is composed by the external layer of the seed (i.e., pericarp, tegmen, and aleurone layer (Figure 1)), translatable in almost 68 million tons of unmanageable material per year, worldwide [2]. Similarly to other cereal species, the kernel surrounding layers are richer in bioactive compounds, minerals, vitamins, dietary fibers, proteins and lipids than the core endosperm, which is characterized by simple carbohydrates and starch granules [3,4] (Table 1). In particular, rice bran has a not-negligible amount of lipids (15–20 g/100 g of RB), where some of the most important bioactive compounds, such as γ-oryzanol, ferulic acid, tocopherol and polyunsaturated fatty acids could be found; thus, for this reason, RB has been used for oil extraction [5]. Despite that, RB is highly sensitive to lipid oxidation due to the rapid activity of lipolytic endogenous enzymes, which means that a thermal stabilization step usually required [6]. However, the first fate of rice bran is the feed formulation industry, losing the opportunity for the recovery of its potential.

Therefore, the study of innovative recovery strategies focused on the valorization of agro-industrial by-products is now an interesting growing sector directed to an improved sustainability and dietary habits of the whole food system. In this work, the strategies for the valorization of rice bran, such as fermentation, air classification and high hydrostatic pressure, were reviewed, with a special focus on their potential modification. Then, its recent nutritional evidences, health-promoting properties and utilization as functional food ingredient will be also summarized.

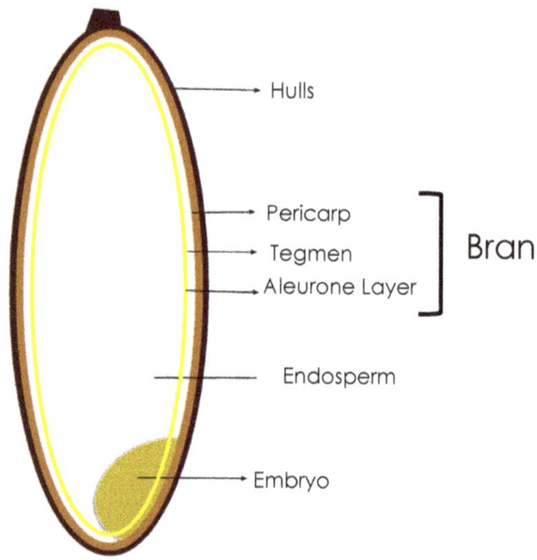

Figure 1. Schematic representation of rice grain, highlighting the bran fraction.

2. Research Criteria for Scientific Papers Selection

The review was organized starting with the study of the most modern techniques used for rice bran treatment. Then, the health-promoting properties related to rice bran were assessed singularly. Finally, the relevance of rice bran as ingredient in the food industry considering its functional properties were reviewed. In this literature review Google Scholar, Web of Science and Scopus research engines were consulted. Key search terms related to rice bran by-products and co-products (i.e., "rice milling by-products", "rice bioactive compounds"), were conjointly searched with the followings: "biotechnology", "fermentation", "emerging technologies", "thermal treatment", "physical treatment", "nutritional value", "health-promoting properties", "safety" and "applications". No data filtering (i.e., year) was applied in order to gain as much knowledge on the topic as possible. The legislation framework information was reported by the European Food Safety Authority (EFSA). Paper selection was performed after a comprehensive study of the results and their discussion as reported by the authors. The output of the latter selection process highlighted 94 original research articles.

3. Composition and Strategies for the Recovery of Rice Bran Potential

The smart recovery of food processing residues is an important aspect for the food industry, which always is in search of value generation. In this context, a great benefit may derive from different branches of science, such as biotechnology, chemistry or physics, applied to the food scenario. There are plenty of innovative techniques applied to produce a high-value ingredient useful for the production of sustainable products by food manufacturers. Following the introduction, rice bran general composition is reported in Table 1. Macro and microelements occurrence depends on several factors including botanical variety, environmental agronomical conditions and processing.

Table 1. Mean proximate composition of rice bran and the content of its main bioactive compounds.

Compound	mg/100 g	References
Carbohydrates	33–42	[7–12]
Proteins	11–16	
Fats	12–20	
Saturated fats	15–20	
Unsaturated fats	75–80	
Dietary fibres (DF)	15–30	
Insoluble DF	13–26	
Soluble DF	1–2.25	
Ash	8–12	
Bioactive Compounds	**mg/g**	
Phenolic acids *	800–1243	
Tocopherols	0.35–0.77	
γ-oryzanol	0.56–1.08	

* as total ferulic acid, the most abundant phenolic acid found in cereal grains (soluble and insoluble forms).

3.1. Bioprocessing of Rice Bran: Solid-State Fermentation and Enzymatic Treatment

Food bioprocessing is that term used to define an operation which combines living microorganism cells or their components to a normal procedure to obtain a different food product. The processes that utilize these technologies are summarized in Table 2. Nowadays, solid-state fermentation (SSF) and the use of specific enzymes are widely studied and applied in food industry, with the principal aim to overall improve the characteristics of the raw materials treated. The most important, from a nutritional point of view, is the fact that the microorganism metabolism or selective enzymes can disrupt the vegetable cell matrix releasing those compounds that present a bioactivity, enhancing their bioavailability in the human organism. Regarding fermentation, both fungi and bacteria can take part to this process. Obviously, the metabolic characteristics of each are different, as well as the resulting raw material modifications. For rice bran fermentation, fungi and yeasts are the mostly used microorganisms. For example, in the study of Abd Razak et al. [13], *Rhizopus oligosporus* and *Monascus purpureus* was used both alone and in combination, for the fermentation of sterilized rice bran. The study analyzed the overall antioxidant capacity (AOC) and bioactive compounds enhancement caused by such treatment. Interestingly, the two microorganisms used in co-culture resulted in a significantly higher AO and phenolic acids content in respect to the non-fermented rice bran. Similar outcomes were obtained by Schmidt et al. [14], where rice bran was fermented with *Rhizopus oryzae (CCT 1217)* over 120 h. They measured AOC with different spectrophotometric and widely used methods, phenolic acids content and the inhibition activity against peroxidase enzyme, which is mainly responsible for the lipid oxidation. The AOC increased in respect to the non-treated rice bran and phenolic compounds, such as ferulic and gallic acids, doubled their initial concentration. This fact can only be attributed to the cell-wall disrupting metabolism of microorganisms, and not to their direct biosynthesis, since fungi cannot produce such secondary metabolites. Similar results were found by Shin et al., in which after the fermentation of black rice bran using *Aspergillus awamori* and *A. oryzae*, they measured a higher soluble phenolic content and radical scavenging activity [15]. Additionally, Oliveira et al. [16] used the same fungus (but a different strain), although they focused the study aim on the lipid compounds modification. Interestingly, the total lipid content of fermented rice bran diminished up to 45% after the fermentation, but the phospholipids increased by 130%, indicating that the microorganism could metabolize such molecules. Moreover, the saturated and unsaturated fatty acids content was altered, the former decreased up to 20% and the latter increased by 5%. Probably ascribed to a combined action of oxidation processes and fungus metabolism. Ryan et al. [17], followed a different but innovative approach to characterize fermented rice bran from three different rice varieties using

Saccharomyces boulardii, a yeast. Employing a metabolomic gas chromatography coupled to mass spectrometry (GC-MS) based technique, they could identify and quantify new bioactive metabolites (i.e., glucitol and unknown disaccharides). Afterword they assessed their bioactivity on normal human peripheral blood lymphocytes (PBL) and malignant human B-cell lymphoma cell lines, resulting in a reduction of the cells growth of the latter ones. Besides phenolic compounds, rice bran polysaccharides (RBPSs) have gained more interest due to their multiple biological benefits. In the study carried out by Liu et al. [18], they fermented a defatted rice bran water extract using the fungus *Grifola frondosa*, aiming to evaluate the composition of the oligosaccharides fraction, AOC and inhibition of nitric oxide (NO) production in macrophages cells. They reported a substantial shift of the polysaccharides molecular weight, from higher to lower, besides the increased AOC and the adjustment of the NO production compared to the non-fermented sample. Furthermore, technologic properties must be considered as equally relevant as nutritional characteristics, since they can firstly affect the food design process. Therefore, Jia et al. [19] focused their study on the effect of defatted rice bran fermentation on soluble dietary fibers, studying their capacity to bind water and oil, which significantly increased after treatment.

Enzymes with specific activities are widely spread in food industry [20]. Their capacity to improve functional and sensory features of food products is accompanied by the fact that their impact on the native composition is very low [21]. Regarding cereal-related products, enzymes are usually used to modify the principal component of that food, such proteins, lipids and carbohydrates [22]. In fact, in the study conducted by Prabhu et al. [23], they treated rice bran with cellulase and xylanase with the aim to solubilize polyphenols with the corresponding improving of the AOC. Similarly, Vallabha et al. [24] treated rice bran with alcalase and endoglucanase, resulting in an increased content of bioactive molecules with no effect on the degradation of B group vitamins after the enzymatic treatment. They also reported an increment of the short chain fatty acids (SCFAs) content in respect to the non-treated rice bran. Moreover, enzymatic processes are usually utilized in combination with other techniques such as fermentation [25], extrusion cooking [26], high hydrostatic pressure (HHP) [27] or micronization [28], leading to an increased extractability of the bioactive component, solubilization of dietary fibers and inhibition of lipid oxidation.

Although these processes might lead to an overall amelioration of the raw material, their biggest disadvantage is that they need long procedures, under specific and strictly controlled conditions to be effective and with the desirable yield. This is only achievable by long term studies performed in tight collaboration between academia and industry. In fact, there is a study published this year [29] in which the authors deeply analyzed the parameters that must be taken into account for the valorization of rice milling side streams under a circular economy view. For example, they reviewed the pre-treatment, the type of fermentation (submerged, solid and semi solid-state, saccharification) useful for the bacterial growth and the subsequent production of valuable compounds like biofuels, lactic acid or biohydrogen.

Considering these treatments, the combined action of enzymes and fermentation can achieve the best results in terms of technological and functional improvements. However, the cost of the processing at industrial scale must be accurately determined. For that reason, the treatment of rice bran using lactic acid bacteria and fungus as fermentation microorganisms seems to be the best condition for obtaining an ingredient with a pleasant organoleptic profile, good nutritional properties and excellent functionalities.

Table 2. Innovative strategies based on bioprocesses applied to rice bran by-product and the novel effects achieved by their application.

Bioprocessing	Effects	References
Solid state fermentation (SSF)	• Increased soluble phenolic acids content; • Increased AOC; • Increased the inhibition of tyrosinase activity;	[13–15]
	• Decreased saturated fatty acids content; • Increased inhibition of peroxidase enzyme; • Increased phospholipids content;	[16]
	• Reduction of the polymerization degree of polysaccharides; • Increased solubilization of soluble dietary fibres; • Potential anti-inflammatory properties; • Higher WHC, OHC, WS and CAC;	[18,19]
Enzymatic treatment (En)	• Increased total phenolic content and AOC; • Higher group B vitamins solubility; • Increased SCFAs (acetic acid and propionic acid) content;	[23,24]
SSF assisted by En	• Increased total phenolic content and AOC.	[25]

Abbreviations: SCFA, short chain fatty acids; SSF, solid state fermentation; AOC, antioxidant capacity; WHC, water holding capacity; OHC, oil holding capacity; WS, water solubility; CAC, cholesterol absorption capacity.

3.2. Thermal and Physical Treatments Applied to Rice Bran

The conscious manipulation of heat has been always used for the transformation of food products in usually, quality-improved foodstuffs. In current years, several new technologies have arisen thanks to their offered potential (Table 3). Extrusion, hot air and far infrared irradiation (FIR) are some of these. For example, Wanyo et al. [30] used hot air and FIR, alone and in combination to assess the effects on the antioxidant properties and bioactives content of rice bran and husk. They reported that the use of FIR technology conferred to rice bran a higher content of bioactive compounds, mainly phenolics, and AOC in respect to the classic hot air treatment. Unfortunately, no mention of the sensorial characteristics was present in the latter study. Furthermore, in the experiment carried out by Dang and Vasanthan [26], they performed a sequential enzyme treatment followed by extrusion, concluding that an increased total soluble pentosan content compared to the to individual or simultaneous treatments. Moreover, Fadel et al. [31], Liu et al. [32] and Dang and Vasanthan [26] applied extrusion techniques at different conditions reporting an increased extraction of arabinoxylans, improved foaming and emulsifying properties and increased soluble fiber content of the extruded bran, respectively. In summary, extrusion is highly studied science field. This is mainly due to the "gold rush" towards meat analogue manufacturing which has occurred in recent years [33]. However, rice bran, and especially rice proteins, are scarcely used as a principal ingredient, since its composition is less suitable to this transformation then other products (i.e., pea protein isolate).

Nevertheless, food science and related disciplines are continuously modernizing to study innovative techniques which may have minimum impact on both products and environment. Physical treatments are those operations based on physics principals applied for many objectives. In the case of cereal products, such rice grain, the properties of porous and fine material (i.e., flour) are the main aspects that must be taken into account. Separation, concentration and physical properties modification of specific food component are the main paths followed by researchers, achieved using dry fractionation, air classification and particle size reduction technologies. These techniques usually result in production of two or more fractions of the material treated, differing at least in granulometry (i.e., fine and coarse fractions). For example, Wang et al. [34], applied pin mill fractionation coupled to an electrostatic separation to defatted rice bran. The mainly outcome authors reported was the dietary fiber enrichment found in coarse fraction of rice bran, remarking the redistribution effect proportioned by this type of technology. Moreover, in the study proposed by Spaggiari et al. [35], authors applied micronization, air-classification and the combination of these techniques in order to obtain different rice bran fractions (fine and coarse), aiming to deeply analyze the lipid profile, with special focus on mono- (MAG) and diacylglycerols (DAG). They found that the lipid content increased significantly in fine fraction, and the MAG, DAG and triacylglycerols content increased by 1.5–5 times in the former fractions. These molecules are particularly interesting to the food industry since they have strong emulsifying potential [36], and find that a "natural source" alternative instead of the additive E-421 (mono- and di-acylglycerol of fatty acids) would be a greater competitive advantage. Nevertheless, more studies have to be conducted in order to optimize this type of treatments, considering the opportunity to combine different fraction with different characteristics, and thus different functions. Furthermore, Silventoinen et al. [37] studied the biochemical and techno-functional modification occurred after the air classification of rice bran, obtaining two different ingredients, a protein-rich ingredient and a fiber-rich Ingredient. These had different microstructural and functional properties, such as protein solubility and colloidal stability, suggesting an elevated applicability in food industry. Moreover, using the HHP technique and different pressures (100, 200 and 500 MPa), Zhu et al. [38] obtained a rice bran with higher WBC and OBC parameters and improved foaming formation and stability, although with a lower least gelation concentration, meaning that the capacity to form a stiff gel requires a lower amount of rice bran. These results stress the importance of the continuous study of food functional properties in order to achieve a higher quality and sustainable ingredients.

Table 3. Thermal and physical treatments applied to rice bran by-product and corresponding outcomes.

Treatment Used	Outcomes	References
Hot-air and FIR	• Higher content of bioactive compounds (tocopherols and phenolic acids); • Higher AOC.	[30]
Extrusion	• Increased arabinoxylans extraction as screw speed increase; • Improvement of the emulsifying and foaming properties.	[31,32]
Extrusion plus enzyme	• Improved dietary fibres solubility;	[26]
Particle size reduction (dry fractionation and micronization)	• Dietary fibres enrichment with modified functional properties; • 1.5–5-fold increment of saturated mono- and di-acylglycerols with potential emulsifying properties in fine RB fraction. • Increased protein content and solubility; • Increased colloidal stability;	[34,35,37]
High Hydrostatic Pressure (HHP)	• Increased ferulic acid extractability; • Inhibition of linoleic acid oxidation; • Higher WBC, OBC, emulsifying stability, improved protein surface hydrophobicity, foaming creation and stabilization.	[27,38]

Abbreviations: RB, rice bran; FIR, far-infrared radiation; WBC, water binding capacity; OBC, oil binding capacity.

4. Rice Bran Health-Promoting Properties

Our knowledge of the relationship between diet and health is becoming stronger. In relation to this, the dietary patterns of developed countries are also changing, showing a growing incidence of many diet-related disorders and diseases (cardiovascular disease (CVD), high cholesterol, diabetes, bowel inflammation, etc.). Rice bran is a by-product rich in relevant bioactive compounds which can play an important role to maintain a healthy status and thus promote a beneficial living style. However, this raw material is currently discarded notwithstanding its potential as a functional ingredient [39] or its potential chemo-preventive and immunomodulatory properties [40–43]. On the other hand, despite the scientifically proven evidence, foods and their constituents should not be considered as medical replacements but as a complementary part which contributes toward a healthy living. However, the scientific evidence-based process is at the base of the applications to specific "health or nutritional claims" [44]. This evaluation procedure is, nowadays, based on European Food Safety Authority (EFSA) opinions regarding a specific application. The health-promoting properties ascribed to rice bran or its valuable compounds are summarized in Figure 2.

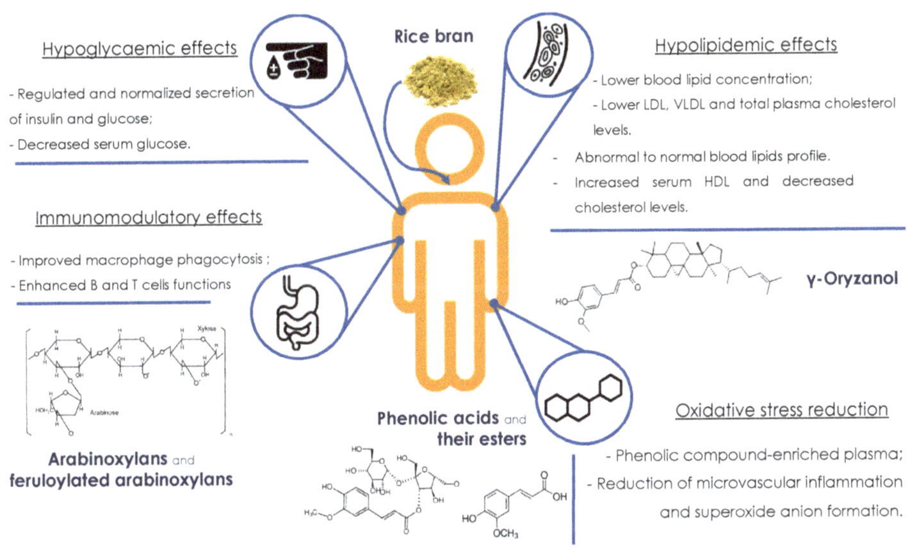

Figure 2. Representation of the health-promoting properties of rice bran, the compounds responsible for that characteristics and the references of the studies highlighting these capacities. Abbreviations: LDL, low density lipoprotein; HDL, high-density lipoprotein; VLDL, very low-density lipoprotein.

4.1. Hypolipidemic Properties of Rice Bran

Recently, an increasing number of studies have analyzed the multiple health-related properties of a rice bran supplemented diet [43]. Besides its high nutrient and bioactive compounds content, rice bran can be administered in different ways, such as stabilized RB, fermented RB, enzymatically treated RB and oil. Each of these products differ from the raw RB, in terms of nutrients and non-nutrients compounds. For this reason, a previous profiling of the product under study is highly recommended in order to define the chemical compounds or mixture of compounds which deliver the desired effects of a previously determined diet.

Health conditions that determine an increased concentration of lipid molecules and cholesterol in blood are a branch of a disease known as hyperlipidemia. These can raise from both genetic conditions and bad dietary habits, which could finally lead to an increased risk of CVD. For this reason, nutrition is the most important variable that must be taken under control for the treatment of such ailment. From this perspective, rice bran and its oil are rich in components which can contrast the accumulation of lipids in blood stream, such as phytosterols, unsaturated fatty acids (FAs) and tocotrienols [41,45]. In fact, the oil recovered from rice bran include glycolipids, phospholipids, free fatty acid and triglycerides with healthy profile, since oleic, linoleic (mono- and polyunsaturated) and palmitic acids (saturated) are the most abundant FAs. In a recent study, Perez-Ternero et al. [46] evaluated the lipid-lowering properties of rice bran enzymatic extract (RBEE) diet supplementation in apolipoprotein E-knockout (ApoE−/−) mice. They reported a higher high-density lipoprotein (HDL) serum value and an increased cholesterol excretion. The same authors reported also other protective properties of RBEE, such as a restored endothelial function, hyperlipidemia prevention, oxidative stress, inflammation and apoptosis reduction in the aorta of ApoE−/− mice [47,48]. Revilla et al. [49] also used RBEE as diet supplementation in male Wistar rats, which resulted in an increased HDL and lower blood cholesterol concentrations in plasma. Besides, Wilson et al. [50] settled up a comparative study in which they evaluated the potential of trans-ferulic acid, γ-oryzanol and rice bran oil supplementation to lower the cholesterol concentration in primates. Among the groups studied, the one fed with γ-oryzanol showed the lower levels of low and very low-density lipoprotein (LDL and VLDL) and plasma cholesterol. Similar outcomes were reported by Ausman et al. [51] and Ha et al., [52], where they fed hypercholesterolemic hamsters and Male Sprague–Dawley rats, respectively, with different amount of rice bran oil. Moreover, Accinni et al. [53] studied the effects of various dietary supplementation, including γ-oryzanol, on the lipid profile of dyslipidemic subjects. Interestingly, results reported by the latter study indicate that the subjects which followed a supplementation diet with γ-oryzanol and niacin recovered a normal lipid blood pattern, better than the other food supplementation studied.

4.2. *Hypoglycaemic Properties of Rice Bran*

The presence of high level of glucose in blood system is a disturb that, nowadays, is extremely widespread in developed country populations, which is mainly associated with bad diet habits and thus could lead to a harmful health status, such as diabetes and hyperglycemia. In this way, vegetable origin food products, such as rice bran, that contain low free sugar and high dietary fiber content can contribute to maintain a lower glycemic index and help to the illness prevention. For example, Son et al. [54] studied the regulation effects of γ-oryzanol on the insulin secretion and glucose concentration in plasma, supplemented to male C57BL/6N mice. The same results were reported in the study of Somsuvra and Ghatak [55], in which adult Wistar rats with a γ-oryzanol supplemented diet had the lower serum glucose content. Moreover, Qureshi et al. [56] expanded the study to humans' volunteers with Type I and II diabetes mellitus. Rice bran was administered as stabilized and as water extract, leading to an increased insulin serum level, decreased glucose and glycosylated hemoglobin concentration.

4.3. *Rice Bran Arabinoxylans: Immunomodulatory Properties*

Recently, the saccharide component of foods has been receiving more attention. The main reason of this is the fact that carbohydrates have never been studied in depth, due to major analytical issues [57]. Despite this, many biological activities have been attributed to molecules including arabinoxylans which have low and high molecular weight. As is widely known, carbohydrates are major components of cereals; in fact, they have the primary function of seed energy storage. In detail, arabinoxylans are composed by xylan backbone (β-(1,4)-D-xylopyranose) linked to α-L-arabinofuranosyl substitutions, which differs among cereal species, conferring different arrangements and thus different functions [58]. In this context, it is crucial to accurately study the chemical composition and potential functionality correlated to arabinoxylans [59]. In fact, a huge effort in terms of research is being made to study the rice bran arabinoxylans-immunomodulatory relationship, which is capable of enhancing the activity of the innate and adaptative responses. These mechanisms are responsible for activating the B-lymphocytes production, our first line of defence from foreign molecules [60,61]. In this way, rice bran is usually enzymatically treated, in order to break high molecular weight arabinoxylans into smaller pieces, in the range of 30–50 Da, producing a product commercially called

BioBran or MGN-3 [62]. These fraction appear to act as an enhancer of natural killer cell, macrophage phagocytosis, B and T cells functions [39]. Ghoneum et al. studied these effects in depth [63–69]. Furthermore, Choi et al. [70] reported similar outcomes after supplementation of the soluble arabinoxylans fraction of wheat bran in mice diet. A proposed theory of the mechanism of action has been proposed [71], stating that the complexity of the structure of the heteropolysaccharides such as galactan, arabinan and β-1,3:1,4- glucan is the basis of the immunomodulatory actions of rice bran arabinoxylans. Other properties (Figure 2) of rice bran arabinoxylans were reported by Salama et al. [72], where the MGN-3 supplementation in patient with chronic hepatitis C virus suppressed the level of viremia. Otherwise, Zheng et al. [73,74] reported protective effects against acute liver injury. In addition, Wang et al. [75] showed an improved anticomplementary activity of DRB under in vitro conditions. Numerous original articles have been published recently which mention the MGN-3 commercial product, although its discussion is not the main focus of these reviews.

4.4. Other Health-Promoting Properties of Rice Bran

Since a wide range of bioactive molecules are present in rice bran, the effects on the oxidative stress are also currently object of study. Justo et al. [76] reported a reduction in microvascular inflammation status in obese Zucker rats which follow a diet supplemented with RBEE. Moreover, Perez-Ternero et al. [77] also studied the supplementation of RBEE in Wistar rats' diet, showing an inactivated superoxide production caused by an increased content of phenolic compounds in blood. Other research has been made focusing on the multiple effects of rice bran supplemented under various forms. For example, fermented brown rice bran using *Aspergillus oryzae*, which induced apoptosis in human acute lymphoblastic leukemia cells [78], or the anti-stress and antifatigue effects delivered by *Saccharomyces cerevisiae* fermented rice bran [79], or otherwise, the improvement of the health condition of mice fed with a high-fat diet including rice bran extract [80]. Moreover, a dietary fibre-rich diet is known to be able to modulate the metabolites produced by human colon microbiota. As reported by Zhang et al. [81], after in vitro gastrointestinal digestion of isolated rice bran dietary fibre, a great probiotic effect on colonic microorganism associated with diabetes was found.

5. Rice Bran Application in Food Industry

The most important step for the recovery of food by-product is the incorporation of such raw material in a classic or novel food product formulation. The primary objective is to add value to the final product, both enhancing nutritional properties or sensorial characteristics, such as color, taste, smell and texture. The latter functions are also important, forming the labelling point of view, since the lack of additives, which are replaced by the recovered material, can lead to the so-called "clean label" status. Rice bran has also another advantage, represented by its suitability for gluten-free product manufacturing. These products usually are poor in term of sensorial quality, but have a high value in market. Moreover, functional properties are another important aspect to be considered for the inclusion of these by-products in a food formulation. These characteristics are reported and summarized in Table 4. RB is a plant material; hence, dietary fibers are the predominant substances composing this product. The latter molecules can contribute to the overall functionality of rice bran, considering its water (WBC) and oil binding (OBC) capacities. Additionally, the protein fraction present in RB (12–16%, [2]) offers potential jellifying, emulsifying and foaming stabilization activities. As stated by Capellini et al., the techno-functional properties of rice bran are interesting for the food industry and they can be improved using specific alcoholic solvents [82]. Moreover, Shi-Wen. et al. [83], reported relevant results regarding the functional properties of rice bran stabilized using different treatments such as microwave and dry heating, with the latter slightly better than the former in relation to the WBC, OBC, emulsifying properties and long-term stability.

Table 4. Functional properties of rice bran reported by scientific original papers.

Functional Property	Outcome	References
Water absorption capacity	• Relevant water retention (>3 g water/g);	
Oil binding capacity	• High oil retention due to insoluble dietary fibres (>3 g oil/g);	
Jellification	• Jellying of hot water solution at 10–18% RB concentration;	
Emulsion stability	• Formation (41–57%) and stabilization of oil in water emulsion (39–74%);	[19,37,38,83,84]
Foaming stabilization	• Low foam formation, but conjointly to other foaming agents is a good stabilizer;	
Protein solubility	• High solubility (around 50%) at acidic (3) and basic (8) pH.	

In fact, it is highly recommendable to stabilize rice bran [85] or to extract the oil before its use as an ingredient, considering its sensibility to oxidation and the subsequent off-flavor formation. This can be easily achieved through the application of thermal and non-thermal techniques, as reported in Section 3.2. Sairam et al. [86] analyzed the addition of different concentration of defatted rice bran (DRB) in bread, aiming to improve the nutritional profile of bread. Total dietary fibres, AOC and the shelf life of bread increased with no repercussions on the sensorial properties. In relation to this, Al-Okbi et al. [87] formulated corn flakes and tortilla chips adding different amounts of rice bran to the original recipe, noticing an organoleptic and rheological amelioration of the final product when the protein content decreased. Premakumari et al. [88] also tried to develop high fiber content ready mixes by replacing the classical cereal flours with a different amount of previously stabilized RB, evaluating their overall acceptability through 20 semi-trained panel members. The main outcome showed that a replacement of at least 25% did not affect the quality of the standard recipe. Moreover, Younas et al. [89] developed cookies using both heat and acid-stabilized RB, optimizing the recipe with a 10% RB substitution. Since RB oil is considered a high added value product (Figure 3), the RB solid exhaust can still be used as a low-fat content ingredient. For example, Charunuch et al. [90] utilized DRB at a different rate for the preparation of extruded breakfast cereal. Meanwhile Alfaro et al. [91] attempted to produce an oil-in-water emulsion using purple and brown rice bran oils, achieving acceptable stability and oxidative protection. However, the use of pretreated rice bran by means of bioprocessing, thermal and physical treatments in food preparation is still scarce, possibly due to the lack of industrial-scale or pilot-scale studies. Moreover, the protein present in rice can be isolated in order to produce ingredients with high protein content, and thus with high functionalities. Regarding this, Zang et al., [92] attempted to improve the emulsification properties of enzymatically-treated rice bran protein, obtaining a more stable emulsion. Nowadays, the topic of plant protein is one which is gaining increasing interest. Therefore, the easy accessibility of rice bran protein represents a relevant resource for the food industry. The proteins recovered from rice bran have specific functionalities, bearing in mind that they are considered hypoallergenic. For this reason, this field is an open door for the food industry, considering that these substances can offer functional and bioactive properties at the same time [93].

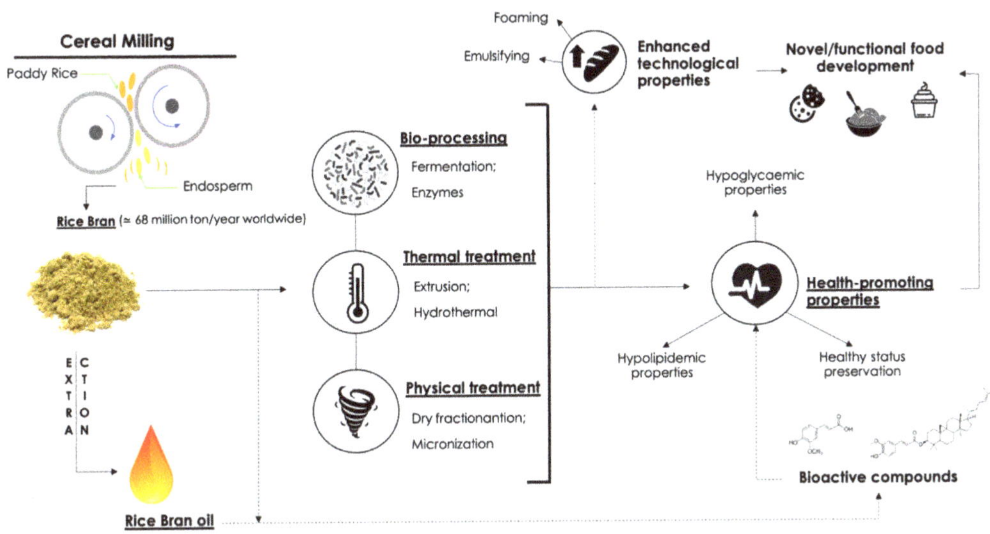

Figure 3. Rice bran production, the treatments applied for its improvement and the nutritional evidences for the formulation of novel or functional food ingredients.

Rice Bran as a Commercial Ingredient

Cereals and their correlated products have always been strong cultural drivers and incubators. Besides, they provide nutrients and energy, constituting the main part of the human diet. Rice grains are typically widespread in Asian and Eastern countries in general, where they are capable of fully exploiting this product. For example, in Japan, there is a special preparation made of fermented rice bran and called "Nukazuke" (Figure 4). Different vegetables are left in this fermented biomass in order to transfer several aromas, tastes and flavors created by the microorganism consortium developed in the rice bran bed [94]; it is a simple pickling. On the contrary, rice in occidental culture is used in different ways, and rice bran has been mostly utilized as a dietary fiber enhancer in industrial food preparations. For example, in baked products such as biscuit [2], leavened pan bread [26] or cakes, the addition of 10 to 20% of stabilized rice bran delivers technological functionality, better nutritional profile and a pleasant texture, without affecting the organoleptic properties compared to the classic recipe. These aspects are extremely important since they can attract the attention of the food industry for the development of new business driven by innovative technologies. In addition, as already mentioned, the oil extracted from rice bran has a high value, although this process might produce "by-by-products", contrasting the idea of thee circular economy and side streams exploitation. However, rice bran oil contains different bioactive compound pro-vitamins, lower saturated fatty acids and polyunsaturated fatty acids, which can be considered as a good nutritional profile. Moreover, the organoleptic characteristics are mild and, if stabilized well, it is a very stable fat [43,52]. The legislation framework is another important aspect when a food product or ingredient comes onto the market. In this case, rice bran is considered as a non-novel food, which is significant for this product. In fact, companies which produce rice bran and other food ingredients are already present and active in the European market (https://www.ricebrantech.com/specialty-ingredients/). They can integrate the entire rice food chain for the recovery of all rice processing side streams.

Figure 4. Representation of the products already present in the market or produced from rice bran by-product.

6. Conclusions and Future Perspective

Rice bran represents a massive industrial by-product with a relevant "hide" potential value. Its compositions, in terms of nutritional features, comprehend a wide range of bioactive compounds which have been studied in recent years. Furthermore, with the aid of the most advanced food-related technologies, the quality parameters of RB can be highly improved, making it a valuable and sustainable resource for healthy diet promotion. In fact, considerable efforts have been made to include this ingredient in widely consumed food products, mainly baked goods. Then, several health-promoting properties suggested the importance of including RB, in its various forms, in the diet. However, the mechanism of action underlying its positive effects should be studied more in depth. In this way, it is important to stress the fact that foods and their components must not be intended as the main instruments with which to struggle with illness. Finally, different types of innovative food products can also be studied, such as novel beverages or dairy products with high nutritional and sensory qualities, helping to achieve a more sustainable food system.

Author Contributions: Conceptualization, M.D.d.C.B. and M.S.; Data Curation, M.S.; Writing—Original Draft Preparation, M.S.; Writing—Review and Editing, M.D.d.C.B., C.D., G.G. and M.S.; Supervision, M.D.d.C.B., C.D. and G.G. All authors have read and agreed to the published version of the manuscript.

Funding: M.S. received a PhD grant by Regione Emilia-Romagna, under the scheme POR-FSE/2016. This research was partially supported by "Nuevos conocimientos para la sostenibilidad del sector cafetero" funded by Consejo Superior de Investigaciones Científicas (CSIC) (201970E117) and "Novel coffee by-product beverages for an optimal health of the brain-gut axis (COFFEE4BGA)" funded by the Ministerio de Ciencia e Innovación (PID2019-111510RB-I00).

Conflicts of Interest: The authors declare no conflict of interest.

References

1. Food and Agriculture Organization of the United Nations: Statistical Databases (FAOSTAT) Production/Crops. *Environ. Sci. Technol.* **2019**, *53*, 11302–11312.
2. Kahlon, T. Rice Bran. In *Fiber Ingredients*; Informa UK Limited: London, UK, 2009; pp. 305–321.

Fritsch, C.; Stäbler, A.; Happel, A.; Márquez, M.A.C.; Aguiló-Aguayo, I.; Abadias, M.; Gallur, M.; Cigognini, I.M.; Montanari, A.; López, M.; et al. Processing, Valorization and Application of Bio-Waste Derived Compounds from Potato, Tomato, Olive and Cereals: A Review. *Sustainability* **2017**, *9*, 1492. [CrossRef]

Schieber, A. Side Streams of Plant Food Processing as a Source of Valuable Compounds: Selected Examples. *Annu. Rev. Food Sci. Technol.* **2017**, *8*, 97–112. [CrossRef] [PubMed]

Gul, K.; Yousuf, B.; Singh, A.K.; Singh, P.; Wani, A.A. Rice bran: Nutritional values and its emerging potential for development of functional food—A review. *Bioact. Carbohydr. Diet. Fibre* **2015**, *6*, 24–30. [CrossRef]

Gopinger, E.; Ziegler, V.; Catalan, A.A.D.S.; Krabbe, E.L.; Elias, M.C.; Xavier, E.G. Whole rice bran stabilization using a short chain organic acid mixture. *J. Stored Prod. Res.* **2015**, *61*, 108–113. [CrossRef]

Ti, H.; Li, Q.; Zhang, R.; Zhang, M.-W.; Deng, Y.; Wei, Z.; Chi, J.; Zhang, Y. Free and bound phenolic profiles and antioxidant activity of milled fractions of different indica rice varieties cultivated in southern China. *Food Chem.* **2014**, *159*, 166–174. [CrossRef]

Elleuch, M.; Bedigian, D.; Roiseux, O.; Besbes, S.; Blecker, C.; Attia, H. Dietary fibre and fibre-rich by-products of food processing: Characterisation, technological functionality and commercial applications: A review. *Food Chem.* **2011**, *124*, 411–421. [CrossRef]

Huang, Y.-P.; Lai, H.-M. Bioactive compounds and antioxidative activity of colored rice bran. *J. Food Drug Anal.* **2016**, *24*, 564–574. [CrossRef]

Chinma, C.E.; Ramakrishnan, Y.; Ilowefah, M.; Hanis-Syazwani, M.; Muhammad, K. REVIEW: Properties of Cereal Brans: A Review. *Cereal Chem. J.* **2015**, *92*, 1–7. [CrossRef]

Ferreira, S.C.; Fernandez, A.M.; de Castillo-Bilbao, M.D. New functional ingredients from agroindustrial by-products for the development of healthy foods. In *Encyclopedia of Food Security and Sustainability*; Elsevier: Amsterdam, The Netherlands, 2018; ISBN 9780128126882.

Razak, D.L.A.; Rashid, N.Y.A.; Jamaluddin, A.; Sharifudin, S.A.; Long, K. Enhancement of phenolic acid content and antioxidant activity of rice bran fermented with Rhizopus oligosporus and Monascus purpureus. *Biocatal. Agric. Biotechnol.* **2015**, *4*, 33–38. [CrossRef]

Schmidt, C.G.; Gonçalves, L.M.; Prietto, L.; Hackbart, H.S.; Furlong, E.B. Antioxidant activity and enzyme inhibition of phenolic acids from fermented rice bran with fungus Rizhopus oryzae. *Food Chem.* **2014**, *146*, 371–377. [CrossRef] [PubMed]

Lim, S.-T.; Kim, S.-M.; Lee, J.H.; Lim, S.-T. Solid-state fermentation of black rice bran with Aspergillus awamori and Aspergillus oryzae: Effects on phenolic acid composition and antioxidant activity of bran extracts. *Food Chem.* **2019**, *272*, 235–241. [CrossRef]

Oliveira, M.D.S.; Feddern, V.; Kupski, L.; Cipolatti, E.P.; Badiale-Furlong, E.; de Souza-Soares, L.A. Changes in lipid, fatty acids and phospholipids composition of whole rice bran after solid-state fungal fermentation. *Bioresour. Technol.* **2011**, *102*, 8335–8338. [CrossRef] [PubMed]

Ryan, E.P.; Heuberger, A.L.; Weir, T.L.; Barnett, B.; Broeckling, C.D.; Prenni, J.E. Rice Bran Fermented with Saccharomyces boulardii Generates Novel Metabolite Profiles with Bioactivity. *J. Agric. Food Chem.* **2011**, *59*, 1862–1870. [CrossRef]

Liu, Q.; Cao, X.; Zhuang, X.; Han, W.; Guo, W.; Xiong, J.; Zhang, X. Rice bran polysaccharides and oligosaccharides modified by Grifola frondosa fermentation: Antioxidant activities and effects on the production of NO. *Food Chem.* **2017**, *223*, 49–53. [CrossRef] [PubMed]

Jia, M.; Chen, J.; Liu, X.; Xie, M.; Nie, S.; Chen, Y.; Xie, J.; Yu, Q. Structural characteristics and functional properties of soluble dietary fiber from defatted rice bran obtained through Trichoderma viride fermentation. *Food Hydrocoll.* **2019**, *94*, 468–474. [CrossRef]

Whitehurst, R.J.; van Oort, M. *Enzymes in Food Technology*, 2nd ed.; Wiley: Hoboken, NJ, USA, 2009; ISBN 9781444309935.

Parrado, J.; Miramontes, E.; Jover, M.; Gutiérrez, J.F.; de Teran, L.C.; Bautista, J. Preparation of a rice bran enzymatic extract with potential use as functional food. *Food Chem.* **2006**, *98*, 742–748. [CrossRef]

Poutanen, K. Enzymes: An important tool in the improvement of the quality of cereal foods. *Trends Food Sci. Technol.* **1997**, *8*, 300–306. [CrossRef]

Prabhu, A.A.; Appukuttan, J.P. Enzymatic processing of pigmented and non-pigmented rice bran on changes in oryzanol, polyphenols and antioxidant activity. *J. Food Sci. Technol.* **2015**, *52*, 6538–6546. [CrossRef]

Vallabha, V.S.; Indira, T.N.; Lakshmi, A.J.; Radha, C.; Tiku, P.K. Enzymatic process of rice bran: A stabilized functional food with nutraceuticals and nutrients. *J. Food Sci. Technol.* **2015**, *52*, 8252–8259. [CrossRef]

Liu, L.; Zhang, R.; Deng, Y.-Y.; Zhang, Y.; Xiao, J.; Huang, F.; Wen, W.; Zhang, M.-W. Fermentation and complex enzyme hydrolysis enhance total phenolics and antioxidant activity of aqueous solution from rice bran pretreated by steaming with α-amylase. *Food Chem.* **2017**, *221*, 636–643. [CrossRef]

Dang, T.T.; Vasanthan, T. Modification of rice bran dietary fiber concentrates using enzyme and extrusion cooking. *Food Hydrocoll.* **2019**, *89*, 773–782. [CrossRef]

Kim, D.; Han, G.D. High hydrostatic pressure treatment combined with enzymes increases the extractability and bioactivity of fermented rice bran. *Innov. Food Sci. Emerg. Technol.* **2012**, *16*, 191–197. [CrossRef]

Wen, Y.; Niu, M.; Zhang, B.; Zhao, S.; Xiong, S. Structural characteristics and functional properties of rice bran dietary fiber modified by enzymatic and enzyme-micronization treatments. *LWT* **2017**, *75*, 344–351. [CrossRef]

Alexandri, M.; López-Gómez, J.P.; Olszewska-Widdrat, A.; Venus, J. Valorising Agro-industrial Wastes within the Circular Bioeconomy Concept: The Case of Defatted Rice Bran with Emphasis on Bioconversion Strategies. *Fermentation* **2020**, *6*, 42. [CrossRef]

29. Wanyo, P.; Meeso, N.; Siriamornpun, S. Effects of different treatments on the antioxidant properties and phenolic compounds rice bran and rice husk. *Food Chem.* **2014**, *157*, 457–463. [CrossRef] [PubMed]
30. Fadel, A.; Plunkett, A.; Ashworth, J.; Mahmoud, A.M.; Ranneh, Y.; el Mohtadi, M.; Li, W. The effect of extrusion screw-speed the water extractability and molecular weight distribution of arabinoxylans from defatted rice bran. *J. Food Sci. Technol.* **2018**, 1201–1206. [CrossRef] [PubMed]
31. Liu, C.; Zhang, Y.; Liu, W.; Wan, J.; Wang, W.; Wu, L.; Zuo, N.; Zhou, Y.; Yin, Z. Preparation, physicochemical and textu properties of texturized rice produce by Improved Extrusion Cooking Technology. *J. Cereal Sci.* **2011**, *54*, 473–480. [CrossRef]
32. Tziva, M.; Negro, S.; Kalfagianni, A.; Hekkert, M. Understanding the protein transition: The rise of plant-based meat substitut *Environ. Innov. Soc. Transit.* **2020**, *35*, 217–231. [CrossRef]
33. Wang, J.; Suo, G.; de Wit, M.; Boom, R.M.; Schutyser, M.A. Dietary fibre enrichment from defatted rice bran by dry fractionati *J. Food Eng.* **2016**, *186*, 50–57. [CrossRef]
34. Spaggiari, M.; Righetti, L.; Folloni, S.; Ranieri, R.; Dall'Asta, C.; Galaverna, G. Impact of air classification, with and witho micronisation, on the lipid component of rice bran (Oryza sativa L.): A focus on mono-, di- and triacylglycerols. *Int. J. Food S Technol.* **2020**, *55*, 2832–2840. [CrossRef]
35. Suman, M.; Silva, G.; Catellani, D.; Bersellini, U.; Caffarra, V.; Careri, M. Determination of food emulsifiers in commercial additiv and food products by liquid chromatography/atmospheric-pressure chemical ionisation mass spectrometry. *J. Chromatogr.* **2009**, *1216*, 3758–3766. [CrossRef]
36. Silventoinen, P.; Rommi, K.; Holopainen-Mantila, U.; Poutanen, K.; Nordlund, E. Biochemical and Techno-Functional Properti of Protein- and Fibre-Rich Hybrid Ingredients Produced by Dry Fractionation from Rice Bran. *Food Bioprocess Technol.* **2019**, 1487–1499. [CrossRef]
37. Zhu, S.M.; Lin, S.L.; Ramaswamy, H.S.; Yu, Y.; Zhang, Q. Enhancement of Functional Properties of Rice Bran Proteins by Hi Pressure Treatment and Their Correlation with Surface Hydrophobicity. *Food Bioprocess Technol.* **2016**, *10*, 317–327. [CrossRef]
38. Sharif, M.K.; Butt, M.S.; Anjum, F.M.; Khan, S.H. Rice Bran: A Novel Functional Ingredient. *Crit. Rev. Food Sci. Nutr.* **2014**, 807–816. [CrossRef] [PubMed]
39. Henderson, A.J.; Ollila, C.A.; Kumar, A.; Borresen, E.C.; Raina, K.; Agarwal, R.; Ryan, E.P. Chemopreventive Properties of Dieta Rice Bran: Current Status and Future Prospects. *Adv. Nutr.* **2012**, *3*, 643–653. [CrossRef] [PubMed]
40. Sohail, M.; Rakha, A.; Butt, M.S.; Iqbal, M.J.; Rashid, S. Rice bran nutraceutics: A comprehensive review. *Crit. Rev. Food Sci. Nu* **2017**, *57*, 3771–3780. [CrossRef] [PubMed]
41. Park, H.-Y.; Lee, K.-W.; Park, H.-Y. Rice bran constituents: Immunomodulatory and therapeutic activities. *Food Funct.* **2017**, 935–943. [CrossRef]
42. Friedman, M. Rice Brans, Rice Bran Oils, and Rice Hulls: Composition, Food and Industrial Uses, and Bioactivities in Huma Animals, and Cells. *J. Agric. Food Chem.* **2013**, *61*, 10626–10641. [CrossRef]
43. General scientific guidance for stakeholders on health claim applications. *EFSA J.* **2016**. [CrossRef]
44. Cicero, A.F.G.; Gaddi, A. Rice bran oil and γ-oryzanol in the treatment of hyperlipoproteinaemias and other conditions. *Phyth Res.* **2001**, *15*, 277–289. [CrossRef] [PubMed]
45. Perez-Ternero, C.; Claro, C.; Parrado, J.; Herrera, M.D.; de Sotomayor, M.A. Rice bran enzymatic extract reduces atheroscleroti plaque development and steatosis in high-fat fed ApoE−/− mice. *Nutrition* **2017**, *37*, 22–29. [CrossRef] [PubMed]
46. Perez-Ternero, C.; Pulgarin, B.B.; de Sotomayor, M.A.; Herrera, M.D. Atherosclerosis-related inflammation and oxidative stre are improved by rice bran enzymatic extract. *J. Funct. Foods* **2016**, *26*, 610–621. [CrossRef]
47. Perez-Ternero, C.; Herrera, M.D.; Laufs, U.; de Sotomayor, M.A.; Werner, C. Food supplementation with rice bran enzymat extract prevents vascular apoptosis and atherogenesis in ApoE−/− mice. *Eur. J. Nutr.* **2015**, *56*, 225–236. [CrossRef] [PubMec
48. Revilla, E.; María, C.S.; Miramontes, E.; Bautista, J.; García-Martínez, A.; Cremades, O.; Cert, R.; Parrado, J.; Torres, M.E. Nutraceutical composition, antioxidant activity and hypocholesterolemic effect of a water-soluble enzymatic extract from ri bran. *Food Res. Int.* **2009**, *42*, 387–393. [CrossRef]
49. Wilson, T.A.; Ausman, L.M.; Lawton, C.W.; Hegsted, D.M.; Nicolosi, R.J. Comparative Cholesterol Lowering Properties Vegetable Oils: Beyond Fatty Acids. *J. Am. Coll. Nutr.* **2000**, *19*, 601–607. [CrossRef] [PubMed]
50. Ausman, L.M.; Rong, N.; Nicolosi, R.J. Hypocholesterolemic effect of physically refined rice bran oil: Studies of cholester metabolism and early atherosclerosis in hypercholesterolemic hamsters. *J. Nutr. Biochem.* **2005**, *16*, 521–529. [CrossRef]
51. Ha, T.-Y.; Han, S.; Kim, S.-R.; Kim, I.-H.; Lee, H.-Y.; Kim, H.-K. Bioactive components in rice bran oil improve lipid profiles in ra fed a high-cholesterol diet. *Nutr. Res.* **2005**, *25*, 597–606. [CrossRef]
52. Accinni, R.; Rosina, M.; Bamonti, F.; Della-Noce, C.; Tonini, A.; Bernacchi, F.; Campolo, J.; Caruso, R.; Novembrino, C.; Ghersi, et al. Effects of combined dietary supplementation on oxidative and inflammatory status in dyslipidemic subjects. *Nutr. Met Cardiovasc. Dis.* **2006**, *16*, 121–127. [CrossRef]
53. Son, M.J.; Rico, C.W.; Nam, S.H.; Kang, M.Y. Effect of Oryzanol and Ferulic Acid on the Glucose Metabolism of Mice Fed with High-Fat Diet. *J. Food Sci.* **2010**, *76*, H7–H10. [CrossRef]
54. Ghatak, S.B.; Panchal, S.S. Anti-diabetic activity of oryzanol and its relationship with the anti-oxidant property. *Int. J. Diabet Dev. Ctries.* **2012**, *32*, 185–192. [CrossRef]
55. Qureshi, A.A.; Sami, S.A.; Khan, F.A. Effects of stabilized rice bran, its soluble and fiber fractions on blood glucose levels ar serum lipid parameters in humans with diabetes mellitus Types I and II. *J. Nutr. Biochem.* **2002**, *13*, 175–187. [CrossRef]

56. Martínez-Montera, C.; Rodríguez-Dodero, M.C.; Guillén-Sánchez, D.A.; Barroso, C.G. Analysis of Low Molecular Weight Carbohydrates in Food and Beverages: A Review. *Chromatographia* **2004**, *59*, 15–30. [CrossRef]
57. Zhang, S.; Li, W.; Smith, C.J.; Musa, H. Cereal-Derived Arabinoxylans as Biological Response Modifiers: Extraction, Molecular Features, and Immune-Stimulating Properties. *Crit. Rev. Food Sci. Nutr.* **2015**, *55*, 1035–1052. [CrossRef] [PubMed]
58. Wang, J.; Bai, J.; Fan, M.; Li, T.; Li, Y.; Qian, H.; Wang, L.; Zhang, H.; Qi, X.; Rao, Z. Cereal-derived arabinoxylans: Structural features and structure–activity correlations. *Trends Food Sci. Technol.* **2020**, *96*, 157–165. [CrossRef]
59. Slack, J. Molecular Biology of the Cell. In *Principles of Tissue Engineering*; Elsevier BV: Amsterdam, The Netherlands, 2014; pp. 127–145.
60. Ooi, S.L.; McMullen, D.; Golombick, T.; Nut, D.; Pak, S.C. Evidence-Based Review of BioBran/MGN-3 Arabinoxylan Compound as a Complementary Therapy for Conventional Cancer Treatment. *Integr. Cancer Ther.* **2017**, *17*, 165–178. [CrossRef] [PubMed]
61. Ghoneum, M.; Matsuura, M. Augmentation of Macrophage Phagocytosis by Modified Arabinoxylan Rice Bran (MGN-3/Biobran). *Int. J. Immunopathol. Pharmacol.* **2004**, *17*, 283–292. [CrossRef] [PubMed]
62. Ghoneum, M.; Gollapudi, S. Modified arabinoxylan rice bran (MGN-3/Biobran) enhances yeast-induced apoptosis in human breast cancer cells in vitro. *Anticancer Res.* **2005**, *25*, 859–870. [PubMed]
63. Elsaid, A.F.; Shaheen, M.; Ghoneum, M. Biobran/MGN-3, an arabinoxylan rice bran, enhances NK cell activity in geriatric subjects: A randomized, double-blind, placebo-controlled clinical trial. *Exp. Ther. Med.* **2018**, *15*, 2313–2320. [CrossRef]
64. Ghoneum, M.; Abedi, S. Enhancement of natural killer cell activity of aged mice by modified arabinoxylan rice bran (MGN-3/Biobran). *J. Pharm. Pharmacol.* **2004**, *56*, 1581–1588. [CrossRef]
65. Ghoneum, M.; Matsuura, M.; Gollapudp, S. Modified Arabinoxylan Rice Bran (Mgn-3/Biobran) Enhances Intracellular Killing of Microbes by Human Phagocytic Cellsin Vitro. *Int. J. Immunopathol. Pharmacol.* **2008**, *21*, 87–95. [CrossRef] [PubMed]
66. Ghoneum, M.; El-Din, N.K.B.; Ali, D.A.; El-Dein, M.A. Modified arabinoxylan from rice bran, MGN-3/biobran, sensitizes metastatic breast cancer cells to paclitaxel in vitro. *Anticancer Res.* **2014**, *34*, 81–87. [PubMed]
67. Gollapudi, S.; Ghoneum, M. MGN-3/Biobran, modified arabinoxylan from rice bran, sensitizes human breast cancer cells to chemotherapeutic agent, daunorubicin. *Cancer Detect. Prev.* **2008**, *32*, 1–6. [CrossRef] [PubMed]
68. Choi, Y.-S.; Lee, J.-K.; Lee, M.-G.; Lee, S.; Jeong, H.-Y.; Kang, H. Splenic T cell and intestinal IgA responses after supplementation of soluble arabinoxylan-enriched wheat bran in mice. *J. Funct. Foods* **2017**, *28*, 246–253. [CrossRef]
69. Fadel, A.; Mahmoud, A.M.; Ashworth, J.J.; Li, W.; Ng, Y.L.; Plunkett, A. Health-related effects and improving extractability of cereal arabinoxylans. *Int. J. Biol. Macromol.* **2018**, *109*, 819–831. [CrossRef] [PubMed]
70. Salama, H.; Medhat, E.; Shaheen, M.; Zekri, A.-R.N.; Darwish, T.; Ghoneum, M. Arabinoxylan rice bran (Biobran) suppresses the viremia level in patients with chronic HCV infection: A randomized trial. *Int. J. Immunopathol. Pharmacol.* **2016**, *29*, 647–653. [CrossRef]
71. Zheng, S.; Sanada, H.; Dohi, H.; Hirai, S.; Egashira, Y. Suppressive Effect of Modified Arabinoxylan from Rice Bran (MGN-3) onD-Galactosamine-Induced IL-18 Expression and Hepatitis in Rats. *Biosci. Biotechnol. Biochem.* **2012**, *76*, 942–946. [CrossRef]
72. Zheng, S.; Sugita, S.; Hirai, S.; Egashira, Y. Protective effect of low molecular fraction of MGN-3, a modified arabinoxylan from rice bran, on acute liver injury by inhibition of NF-κB and JNK/MAPK expression. *Int. Immunopharmacol.* **2012**, *14*, 764–769. [CrossRef]
73. Wang, L.; Zhang, H.; Zhang, X.; Chen, Z. Purification and identification of a novel heteropolysaccharide RBPS2a with anti-complementary activity from defatted rice bran. *Food Chem.* **2008**, *110*, 150–155. [CrossRef]
74. Justo, M.; Claro, C.; Vila, E.; Herrera, M.; Rodríguez-Rodríguez, R. Microvascular disorders in obese Zucker rats are restored by a rice bran diet. *Nutr. Metab. Cardiovasc. Dis.* **2014**, *24*, 524–531. [CrossRef]
75. Perez-Ternero, C.; Macià, A.; de Sotomayor, M.A.; Parrado, J.; Motilva, M.J.; Herrera, M.-D. Bioavailability of the ferulic acid-derived phenolic compounds of a rice bran enzymatic extract and their activity against superoxide production. *Food Funct.* **2017**, *8*, 2165–2174. [CrossRef] [PubMed]
76. Horie, Y.; Nemoto, H.; Itoh, M.; Kosaka, H.; Morita, K. Fermented Brown Rice Extract Causes Apoptotic Death of Human Acute Lymphoblastic Leukemia Cells via Death Receptor Pathway. *Appl. Biochem. Biotechnol.* **2016**, *178*, 1599–1611. [CrossRef] [PubMed]
77. Kim, K.M.; Yu, K.W.; Kang, D.H.; Suh, H.J. Anti-stress and anti-fatigue effect of fermented rice bran. *Phytother. Res.* **2002**, *16*, 700–702. [CrossRef]
78. Duansak, N.; Piyabhan, P.; Srisawat, U.; Naowaboot, J.; Lerdvuthisopon, N.; Schmid-Schönbein, G. The Effect of Rice Bran Extract on Arterial Blood Pressure, Hepatic Steatosis, and Inflammation in Mice Fed with a High-Fat Diet. *J. Nutr. Metab.* **2020**, *2020*, 1–9. [CrossRef] [PubMed]
79. Zhang, X.; Zhang, M.; Dong, L.; Jia, X.; Liu, L.; Ma, Y.; Huang, F.; Zhang, R. Phytochemical Profile, Bioactivity, and Prebiotic Potential of Bound Phenolics Released from Rice Bran Dietary Fiber during in Vitro Gastrointestinal Digestion and Colonic Fermentation. *J. Agric. Food Chem.* **2019**, *67*, 12796–12805. [CrossRef]
80. Capellini, M.C.; Giacomini, V.; Cuevas, M.S.; Rodrigues, C.E. Rice bran oil extraction using alcoholic solvents: Physicochemical characterization of oil and protein fraction functionality. *Ind. Crop. Prod.* **2017**, *104*, 133–143. [CrossRef]
81. Lv, S.-W.; He, L.-Y.; Sun, L.-H. Effect of different stabilisation treatments on preparation and functional properties of rice bran proteins. *Czech J. Food Sci.* **2018**, *36*, 57–65. [CrossRef]
82. Capellini, M.C.; Novais, J.S.; Monteiro, R.F.; Veiga, B.Q.; Osiro, D.; Rodrigues, C.E.D.C. Thermal, structural and functional properties of rice bran defatted with alcoholic solvents. *J. Cereal Sci.* **2020**, *95*, 103067. [CrossRef]

83. Liu, Y.Q.; Strappe, P.; Zhou, Z.; Blanchard, C. Impact on the nutritional attributes of rice bran following various stabilization procedures. *Crit. Rev. Food Sci. Nutr.* **2018**, *59*, 2458–2466. [CrossRef]
84. Sairam, S.; Krishna, A.G.G.; Urooj, A. Physico-chemical characteristics of defatted rice bran and its utilization in a bakery produc *J. Food Sci. Technol.* **2011**, *48*, 478–483. [CrossRef]
85. Al-Okbi, S.Y.; Hussein, A.M.; Hamed, I.M.; Mohamed, D.A.; Helal, A.M. Chemical, Rheological, Sensorial and Functiona Properties of Gelatinized Corn- Rice Bran Flour Composite Corn Flakes and Tortilla Chips. *J. Food Process. Preserv.* **2012**, *38*, 83–8 [CrossRef]
86. Mohan, J. Development and Acceptability of Fibre Enriched Ready Mixes. *Int. J. Pure Appl. Sci. Technol.* **2012**, *9*, 74–83.
87. Shaheen, M.; Ahmad, I.; Anjum, F.M. Effect of processed rice bran supplementation on the quality of chapaties. *Thai J. Agric. Sc* **2012**, *45*, 241–247.
88. Charunuch, C.; Limsangouan, N.; Prasert, W.; Butsuwan, P. Optimization of Extrusion Conditions for Functional Ready-to-Ea Breakfast Cereal. *Food Sci. Technol. Res.* **2011**, *17*, 415–422. [CrossRef]
89. Alfaro, L.; Zhang, J.; Chouljenko, A.; Scott, R.; Xu, Z.; Bankston, D.; Bechtel, P.J.; Sathivel, S. Development and characterization o emulsions containing purple rice bran and brown rice oils. *J. Food Process. Preserv.* **2016**, *41*, e13149. [CrossRef]
90. Zang, X.; Yue, C.; Wang, Y.; Shao, M.-L.; Yu, G. Effect of limited enzymatic hydrolysis on the structure and emulsifying propertie of rice bran protein. *J. Cereal Sci.* **2019**, *85*, 168–174. [CrossRef]
91. Zaky, A.A.; El-Aty, A.M.A.; Ma, A.; Jia, Y. An overview on antioxidant peptides from rice bran proteins: Extraction, identificatior and applications. *Crit. Rev. Food Sci. Nutr.* **2020**, 1–13. [CrossRef]
92. Ono, H.; Nishio, S.; Tsurii, J.; Kawamoto, T.; Sonomoto, K.; Nakayama, J. Monitoring of the microbiota profile in nukadoko, naturally fermented rice bran bed for pickling vegetables. *J. Biosci. Bioeng.* **2014**, *118*, 520–525. [CrossRef]
93. Sharif, M.K.; Butt, M.S.; Anjum, F.M.; Nawaz, H. Preparation of Fiber and Mineral Enriched Defatted Rice Bran Supplemente Cookies. *Pak. J. Nutr.* **2009**, *8*, 571–577. [CrossRef]
94. Ajmal, M.; Butt, M.; Sharif, K.; Nasir, M.; Nadeem, M. Preparation of Fiber and Mineral Enriched Pan Bread by Using Defattec Rice Bran. *Int. J. Food Prop.* **2006**, *9*, 623–636. [CrossRef]

foods

MDPI

Review

Exploitation of Agro-Industrial Waste as Potential Source of Bioactive Compounds for Aquaculture

Nayely Leyva-López [1,2], Cynthia E. Lizárraga-Velázquez [2], Crisantema Hernández [2,*] and Erika Y. Sánchez-Gutiérrez [2]

1 Cátedras CONACYT-Centro de Investigación en Alimentación y Desarrollo, A.C. (Food and Development Research Center), Unidad Mazatlán. Av. Sábalo Cerritos S/N, Mazatlán 82112, Sinaloa, Mexico; nayely.leyva@ciad.mx

2 Centro de Investigación en Alimentación y Desarrollo, A.C. (Food and Development Research Center), Unidad Mazatlán. Av. Sábalo Cerritos S/N, Mazatlán 82112, Sinaloa, Mexico; cesmeralda.lizarraga@gmail.com (C.E.L.-V.); erika.sanchez@ciad.mx (E.Y.S.-G.)

* Correspondence: chernandez@ciad.mx

Received: 26 May 2020; Accepted: 18 June 2020; Published: 28 June 2020

Abstract: The agroindustry generates a large amount of waste. In postharvest, food losses can reach up to 50%. This waste represents a source of contamination of soil, air, and bodies of water. This represents a problem for the environment as well as for public health. However, this waste is an important source of bioactive compounds, such as phenolic compounds, terpenes, and β-glucans, among others. Several biological activities have been attributed to these compounds; for example, antioxidant, antimicrobial, gut microbiota, and immune system modulators. These properties have been associated with improvements in health. Recently, the approach of using these bioactive compounds as food additives for aquaculture have been addressed, where it is sought that organisms, in addition to growing, preserve their health and become disease resistant. The exploitation of agro-industrial waste as a source of bioactive compounds for aquaculture has a triple objective—to provide added value to production chains, reduce pollution, and improve the well-being of organisms through nutrition. However, to make use of the waste, it is necessary to revalue them, mainly by determining their biological effects in aquaculture organisms. The composition of bioactive compounds of agro-industrial wastes, their biological properties, and their application in aquaculture will be addressed here.

Keywords: food by-products; biological activities; phenolic compounds; antioxidant; immunostimulants; prebiotics; fish

1. Introduction

Population growth and urbanization have increased the demand for processed foods. This has led to the development of food industries to meet the needs of consumers. In particular, agricultural industries generate a large amount of waste during the collection, storage, transport, and processing of raw materials [1]. This represents an environmental pollution problem due to the waste being mainly made up of organic matter. Organic matter represents a source of bioactive compounds. A bioactive compound is a substance that has a biological activity. In a broader sense, it is a substance that has an effect or can trigger a physiological response on a living organism. The effect may be negative or positive depending of the chemical structure, the dose, and the bioavailability of the substance [2]. However, bioactive compounds are widely recognized for promoting health benefits [3].

In Mexico, the processing industries of fruits (lemon, pears, tomato, apples, papaya, pineapple, banana, and oranges), cereals (corn), and vegetables (beans, cabbage, carrots, lettuce, and potatoes) generate around 76 million tons annually of waste [4]. The husks, bark, seeds, and pomace represent the largest amount of waste, which are made up of a wide variety of bioactive compounds with multiple biological properties such as antioxidant, immunostimulant, antimicrobial, anticancer and prebiotic.

The exploitation of waste for the development of food products with added value could make possible the generation of additional profits. The agricultural industry, in collaboration with the scientific community, has directed efforts toward the design of appropriate methods for the extraction and purification of bioactive compounds from waste for the development of functional food. It is currently known that the cost of technologies for the purification of bioactive compounds exceed reprocessing costs, so the full use of waste with functional properties as additives could lead the food industry to reduce residues and increase its profitability [5].

Currently, the aquaculture sector is in need to find low-cost natural compounds as food additives, which might promote well-being and preserve the nutritional quality of farming organisms, without compromising the environment and the health of consumers. Therefore, the use of plant residues as additives represents a promising alternative. However, these residues require appropriate characterization, since some studies report that the use of phytochemicals might exert anti-nutritional effects [6,7].

2. Bioactive Compounds from Agro-Industrial Waste

There is a wide variety of bioactive compounds found in residues derived from the cultivation and processing of agricultural products. Any part of plants, such as husk, seeds, leaves, roots, and stems, can be considered as a source of bioactive compounds [8]. Some of the most important bioactive groups are briefly described next.

2.1. Phenolic Compounds

Phenolic compounds (PCs) are secondary metabolites of plants that are usually esterified or glycosylated [9] and are mainly composed of an aromatic ring (hydrophobic domain) and one or more hydroxyl groups (hydrophilic domain) attached to it. Within this group of compounds are phenolic acids (hydroxycinnamic and hydroxybenzoic), flavonoids, coumarins, xanthones, chalcones, stilbenes, lignins, and lignans (Figure 1a) [10].

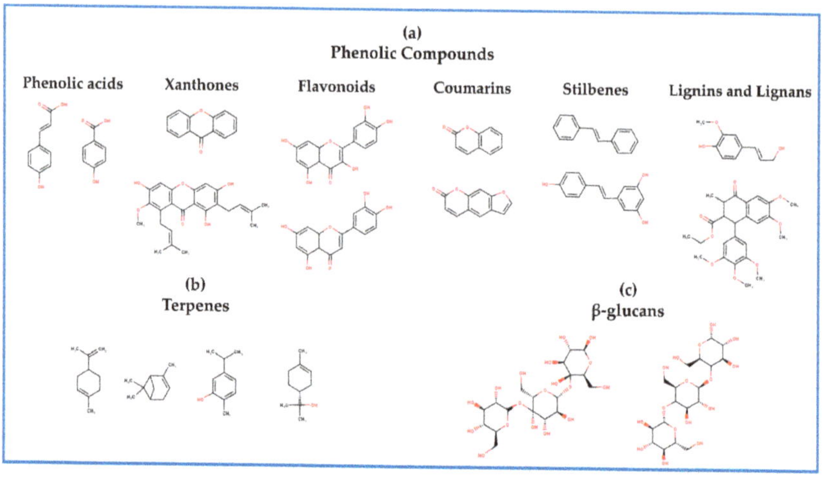

Figure 1. Chemical structure of (**a**) phenolic compounds, (**b**) terpenes, and (**c**) β-glucans.

Phenolic compounds are known for having several biological properties, such as antioxidants [11], immunostimulants [12], and microbiota modulators [13]. Besides, they are also recognized for their antibacterial [14], antiparasitic [15], antiviral [16], anti-inflammatory [17], anticancer [18], and antihypertensives [19] effects. The biological properties of PCs, especially the antioxidant, are related to their chemical-structural characteristics. The number and position of hydroxyl groups, the presence of double bonds, and the ability to delocalize electrons determine the ability of PCs to scavenge free radicals and donate hydrogen atoms [20]. Other mechanisms by which PCs can exert their biological activity is by interacting with components of the cell membrane, enzymes, and transcription factors, as well as receptors [21]. For example, PCs provide antioxidant protection to membrane through the interaction of their hydrophilic and hydrophobic domains with the polar head and non-polar chain of lipids bilayer [21]. Flavonoids exert antiallergenic and anticarcinogenic activities by interacting with membrane raft-associated proteins [22] and decrease superoxide anion production in vascular cells through the inhibition of the translocation of $p47^{phox}$ nicotinamide adenine dinucleotide phosphate (NADPH) oxidase subunit in the endothelial cell membrane [19]. Moreover, PCs exert indirect antioxidant and anti-inflammatory activities by activation of Nrf2 and inhibition of NF-κB translocation into the nucleus, respectively [23].

2.2. Terpenes

Terpenes are compounds formed by isoprene units (C_5H_8). These compounds are classified according to the number of isoprene units condensed [24]. Within the group of terpenes are essential oils and carotenoids, which are characterized by their antioxidant activity. Also, essential oils are known to exert microbicidal effects [25,26].

Terpenes biological activities are related to their chemical structure (Figure 1b). For instance, the antioxidant activity of essential oils is mainly due to the presence of phenolic type components, such as thymol and carvacrol; therefore, the antioxidant action mechanisms of these terpenes are similar to those previously mentioned for PCs [27]. On the other hand, the antimicrobial effect of essential oils is related to their hydrophobicity. This property allows essential oils to cross the cell wall and cytoplasmic membrane and disorganize the structure of its components. Besides, essential oils are capable of inhibiting enzymes of energy regulation and synthesis of structural components [28].

2.3. Dietary Fiber (β-glucans)

Dietary fiber is composed of polymers of three or more carbohydrate units that are resistant to the activity of endogenous digestive enzymes, and therefore cannot be hydrolyzed or absorbed by the small intestine [29]. Fiber is classified as insoluble and soluble fiber. Soluble fibers, such as β-glucans, fructooligosaccharides, galactooligosaccharides, and some pectins, are fermented by the intestinal microbiota and give rise to short-chain fatty acids (acetate, propionate, and butyrate) [30].

Particularly in aquaculture, β-glucans (Figure 1c) are recognized for their immunostimulatory activity [31]. These compounds are polysaccharides made up of glucose units linked by glycosidic bonds. These are found as components of the cell wall of plants and yeasts mainly, but also in some species of algae and fungi. The most relevant β-glucans are β-1,3 and β-1,6 [32,33]. The immunostimulatory activity of β-glucans depends on their recognition and binding to membrane receptors (for example, dectin-1, and CR3). Besides, the degree of polymerization, the degree and type of branching, and the structural conformation of β-glucans affect their interaction with receptors [34].

2.4. Glucosinolates

The glucosinolates are glycosides formed by a β-D-glucopyranose residue linked to a (Z)-N-hydroximinosulfate ester by sulfur bridges, and an amino acid derivative radical. These compounds are found in all species belonging to the *Brassica* family, such as canola, broccoli, arugula, and mustard [35]. Glucosinolates can be classified based on their amino acid precursor into aliphatic, aromatic, and indole [36,37].

Glucosinolates and the products derived from their degradation (isothiocyanates) show antioxidant, anticancer and antibacterial activity. These compounds act as indirect antioxidants because they are capable of modulating the activity of xenobiotic-metabolizing enzymes (Phase I and Phase II), which triggers the long-lasting antioxidant reactions [38]. On the other hand, the bactericidal activity of the products of the metabolism of glucosinolates has been related to the inhibition of intracellular enzymes responsible for ATP synthesis in pathogenic bacteria [39,40].

2.5. Saponins

Saponins are amphipathic molecules composed of sugar residues linked to a system of polycyclic rings (sterols and triterpenes) through glycosidic bonds [41]. These compounds are present in plant products, such as agave or legumes [42,43].

Saponins have immunostimulatory effects [44]. The structural characteristic associated with this activity is the presence of an aldehyde group at position C19 and C4 of the aglycone [45]. Besides, saponins exert microbiota modulating effect, which is related to their antimicrobial activity. Furthermore, saponins can dissociate the cell membrane, and therefore, the flow of extracellular and intracellular components is enabled [46]. The effectiveness of saponins is enhanced against Gram-positive bacteria, while Gram-negative bacteria are more resistant, possibly due to the presence of the double lipid membrane in the latter [47].

Despite the beneficial properties attributed to bioactive compounds, they might possess anti-nutritional effects due to inhibition of the digestive protease activity and formation of complexes with proteins [48,49]. Since bioactive compounds might exert beneficial effects on organisms of importance for aquaculture, their use as food additives has been explored. Nevertheless, the effect of these compounds on the metabolism and growth of species is still to be understood.

3. Biological Properties and Mode of Action of Bioactive Compounds

3.1. Antioxidant Activity

Free radicals are atoms or molecules that have a missing electron in the last orbital, which gives them instability and high reactivity. Free radicals reach balance by receiving electrons from other molecules, such as carbohydrates, proteins, lipids, and nucleic acids [50]. These reactive molecules are produced during normal cellular metabolism, some examples are superoxide anion (O^{2-}), hydroxyl radical (OH^-) and hydroperoxyl radical (HO^{2-}) [51]. An excess in the levels of free radicals can start harmful effects on important macromolecules, like lipids, proteins and nucleic acids [52]. The lipid peroxidation is caused by free radicals. This process increases the production of free radicals and leads to the formation of aldehydes such as malondyaldehyde (MDA) and 4-hydroxy-2-nonenal (HNE) (Figure 2a), which are characterized by their cytotoxic and mutagenic effects [52,53]. Lipid peroxidation and other cell damages can be prevented with antioxidants.

Antioxidants are substances capable of neutralizing or reducing the deterioration caused by free radicals [54]. The antioxidant activity can be exerted by directly donating electrons to free radicals to stabilize them or regulating the activity of transcription factors, such as the nuclear factor enhancing the kappa light chains of activated B cells (NF-κB) and the nuclear factor derived from erythroid 2 (Nrf2). These factors participate in the regulation of gene expression of detoxifying and antioxidant enzymes such as superoxide dismutase (SOD), catalase (CAT), and glutathione peroxidase (GPx) (Figure 2b) [55,56].

Bioactive compounds act as both direct and indirect antioxidants due to the presence of OH groups, double bond, carbonyl groups, and aromatic rings in their structures. These compounds can regulate the Nrf2 activation by Keap-1-dependent and Keap-1-independent pathways [57]. Keap-1-dependent requires strong electrophiles disrupting the Keap1-Nrf2 complex. Keap1-independent involves the phosphorylation of Nrf2 by protein kinases (ERK, JNK, PKC, p38 MAPK, and Akt) [57]. Both pathways promote the Nrf2 translocation to nucleus and the binding of this factor to sMaf protein and the antioxidant response element (ARE). This initiates the transcription of antioxidant enzymes (Figure 2c), such as SOD, CAT and GPx. Gallic and caffeic acids activate Nrf2 by induction of the post-translational phosphorylation of ERK [22,57], while β-glucans do it via p38 MAPK signaling [58]. Electron rich flavonoids such as epigallocatechin gallate, quercetin, and morin have the ability to form stabilized electrophiles and act as

Michael reaction acceptors and, thus, can modify cysteine residues of Keap1 [59]. Carotenoids such as lycopene and adonixanthin can also activate Nrf2, although the activation pathway is still unknown [60].

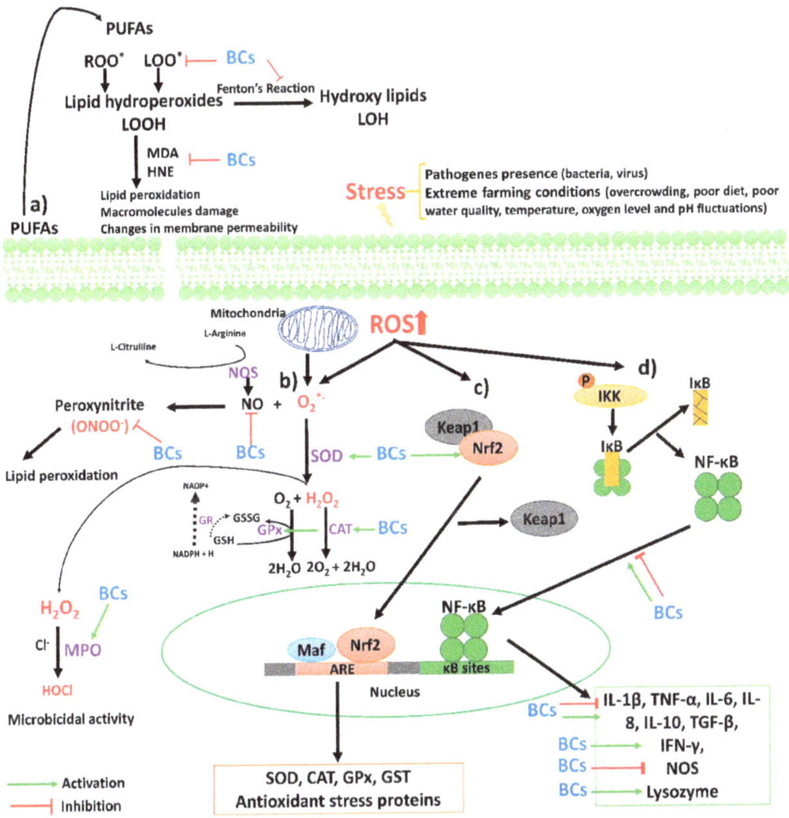

Figure 2. Graphical representation of the mechanism of action of bioactive compounds on the antioxidant and immune response. (**a**) Lipid peroxidation chain reaction, (**b**) antioxidant enzymes reaction, (**c**) Nrf2 pathway associated to the antioxidant response, and (**d**) NF-κB pathway associated to the immune response. Abbreviations: ARE—antioxidant response element; BCs—bioactive compounds; CAT—catalase; GPx—glutathione peroxidase; GR—glutathione reductase; GSH—glutathione; GSSG—oxidized glutathione; GST—glutathione transferase; HNE—4-hydroxynonenal; HOCl—hypochlorous acid; IFN-γ—interferon-gamma; IkB—inhibitor protein of nuclear factor kappa-light chain-enhancer of activated B cells; IKK—kinase complex; IL—interleukin; Keap1—Kelch-like ECH-associated protein 1; LOO*—lipid hydroperoxyl radical; Maf—musculoaponeurotic fibrosarcoma; MDA—malondialdehyde; MPO—myeloperoxidase; NADP+—nicotinamide adenine dinucleotide phosphate; NADPH—reduced form of NADP; NF-κB—nuclear factor kappa-light chain-enhancer of activated B cells; NOS—nitric oxide synthase; Nrf2—NF-E2-related factor 2; PUFAs—polyunsaturated fatty acids; ROO*—peroxyl radical; SOD—superoxide dismutase; TGF-β—transforming growth factor-beta; TNF-α—tumor necrosis factor-alpha.

The NF-κB activation occurs by phosphorylation of protein kinases (PI3K, PKC, JNK, and ERK), which induce phosphorylation of IKKα/β and leads to IκBα phosphorylation. Afterward, the NF-κB is translocated to the nucleus for the transcriptional regulation of antioxidant genes (Figure 2d) [61]. The high electrophilicity of Michael acceptors of PCs also allows to modify IKK cysteine residues and hence, the nuclear translocation of NF-κB [61,62]. However, PCs also can act as anti-inflammatory agents and avoid the NF-κB activation by inhibition of the phosphorylation of IκB-α, ERK, JNK, p38, and MAPK [62]. In this sense, lycopene also can inactivate the NF-κB translocation [63].

The gene expression and activity of antioxidants enzymes (SOD, CAT, GPx) and the levels of lipid peroxidation mediators—namely, malonaldehyde (MDA) are used as markers to determine if a food additive can exert a stimulating action on the antioxidant system and reduce oxidative stress in the organisms [64,65].

3.2. Immunostimulant Activity

Immunostimulants, also known as adjuvants or immunomodulators, are compounds or substances that promote the response of the immune system, which might generate resistance of the organisms to the presence of pathogens [66]. Immunostimulants can be obtained from animal, vegetable, and bacterial sources, as wells as from algae, nutritional factors and hormones, or cytokines [67]. Generally evaluated biomarkers used to define the immunostimulant effect of a substance are:

(1) The increase in the enzymatic activity of lysozyme and myeloperoxidase (MPO). Lysozyme exerts its microbicidal action by lysis of peptidoglycans, components of the cell wall of Gram-positive bacteria [68], while MPO catalyzes the formation of hypochlorous, hypobromous and hypothiocyanite acids [69].

(2) Increase in respiratory burst. When phagocytic cells, such as neutrophils and macrophages, respond to the presence of a pathogen, they trigger the action of NADPH oxidase. This generates superoxide anion (O^{2-}). The measurement of this radical by the nitroblue tetrazolium (NBT) reduction method has been considered as an indicator of the phagocytic capacity of the cells of the immune system [70].

(3) Increase in the number of red and white cells. The cell count is a measure used to evaluate the effect of some possible immunostimulants on the health of organisms. A reduction in the count of red cells (erythrocytes) implies that the substance is exerting collateral damage (anemia) in the body. An increase in the number of white cells (leukocytes) indicates a greater response of the immune system to a possible infectious agent. Other blood cell indicators are neutrophils count, hematocrit, level of hemoglobin, etc. [66].

(4) Other immunological parameters evaluated are complement components, such as soluble proteins, enzymes, and receptors that act in signaling processes, opsonization of pathogenic microbes, phagocytosis and microbial destruction [71]. The concentration of immunoglobulins (Ig) and the level of protein are also frequently evaluated as immunological parameters [66]. Melanomacrophage centers (MMCs), pigmented phagocytic cells (melanin) that act as a rapid response to the presence of an infection, and cytokine levels, such as interleukin-1 (IL-1), IL-6, and interferon-gamma (IFN-γ) are also considered markers of the immune response in fish [72].

The NF-κB factor plays an extremely important role in the immune response of organisms. It is responsible for the expression of inflammatory genes, namely cytokines and enzymes (nitric oxide synthase, NOS). Furthermore, NF-κB is activated under stress conditions, such as oxidative agents or pathogens presence (Figure 2d). These stressors provoke the phosphorylation of IκB accompanied by the ubiquitination and degradation of the protein. The IκB degradation releases the NF-κB and allows it to enter the nucleus and activate the expression of immune-related genes [73]. Therefore, the NF-κB modulation is a key factor to evaluate the immunomodulatory effect of bioactive compounds.

It has been proposed that PCs, such as flavonoids, could exert anti-inflammatory effects by inhibiting the action of the NF-κB factor. For instance, the -OH groups attached to the C3′ and C4′ in B ring present in (−)-epicatechin may interact with the nuclear fractions p50 and Rel A of NF-κB, which prevents the binding of the factor with the specific DNA-kB sites, hence blocking the expression of cytokine-related genes. [21]. In contrast, it has been demonstrated that quercetin, a flavonoid, increases IFN-γ secretion while reduces IL-4 levels in blood mononuclear cells [74]. An up-regulation of IFN-γ has been associated to a better adaptive immune response in fish [75]. Due to these inconsistencies, more research is needed on bioactive compounds to obtain conclusive data. Monoterpenes, such as limonene, have been reported to inhibit the phosphorylation of IkB and therefore block the translocation of the factor NF-κB to the nucleus [76]. On the other hand, β-glucans exert their immunomodulatory effect by promoting the activation of NF-κB via recognition and binding by the receptors dectin-1 and CR3, among others. Dectin-1 recognizes glucans with both β-(1–3) and β-(1–6) bonds, therefore the linking strength depends on the size of the molecule and branching degree. Through this recognition, the NF-κB is activated to induce cytokine and chemokine synthesis [34]. In fish, a member of the C-type lectin receptor, different from dectin-1, might be responsible of the β-glucan recognition. Two C-type lectin domain-encoding genes, named *clec4c* and *sclra*, were identified in primary macrophages of common carp (*Cyprinus carpio* L.) [77].

3.3. Intestinal Microbiota Modulation

The modulation of the intestinal microbiota or the induction of changes in the composition of the host's microbiota is achieved through the use of probiotics, prebiotics, and synbiotics [78]. A prebiotic is defined as a non-digestible compound that, through its metabolism by microorganisms, modulates the composition or activity of the intestinal microbiota, and confers a beneficial physiological effect on the host [79]. Generally, the modulation of the microbiota is carried out for the benefit of the host to increase the abundance of beneficial bacteria and inhibit the growth of pathogenic bacteria. The latter can be achieved by selecting additives and/or functional ingredients, that once incorporated into the food might exert the activity.

The immune modulatory effect of β-glucans is strongly related to their chemical structure and depends on the type of branching, the degree of ramifications, solubility, molecular weight, tertiary structure, polymer charge, and solution conformation (triple or single helix or random coil) [80]. For instance, soluble β-glucans are fermented by the gut microbiota in the large intestine, such as populations of bifidobacteria and lactobacilli, which produce cell-associated glycosidases. After fermentation, β-glucans are metabolized to produce short-chain fatty acids, such as acetic, propionic and butyric acids [81]. These short-chain fatty acids have been reported to possess biological activity, such as reducing cholesterol levels in humans [82]. The exact molecular mechanism by which β-glucans affect gut microbiota is still not clear [83]; therefore, more research is needed in this topic.

Dietary fibers are the plant bioactive compounds more studied due to their ability to modify the intestinal microbiota (prebiotic effect) [79]. Some PCs that are not absorbed in the intestine might be metabolized by intestinal bacteria, acting as modulators of the microbiota [84]. The precise mechanism by which PCs exert their microbiota modulation effect still remains to be elucidated. Nevertheless, there are reports that show that gut microbiota transforms PCs that are linked to glycosides [85]. This transformation occurs through a series of reactions of hydroxylation, methylation, dehydrogenation, isomerization, glycosylation, etc. [13]. For instance, epigallocatechin gallate, from tea, was converted to its derivates gallic acid and epigallocatechin after 36 h of anaerobic fermentation in vitro inoculated with fecal slurry from healthy women. Furthermore, PCs from green tea, oolong tea, and black tea significantly increased population of bacteria belonging to genera *Bifidobacterium* and *Lactobacillus-Enterococcus* spp. and suppressed the growth of *Bacteroides-Prevotella* and *Clostridium histolyticum* [86].

The modulatory effect of bacteria by terpenes has been widely associated to the antibacterial properties of these bioactive compounds. Terpenes can pass through the cell membrane of bacteria and cellular organelles due to their hydrophobic properties and therefore disarrange the structure of the phospholipidic bilayer and increase permeability. This causes the leakage of relevant molecules and ions in the bacteria [26]. Furthermore, by destabilizing the membrane structure of cellular organelles, such as mitochondria and endoplasmic reticulum, terpenes inhibit enzymatic reactions responsible of the energy metabolism, as well as the synthesis of structural macromolecules. Besides, it is proposed that terpenes, namely essential oils, exert a higher antibacterial effect on Gram-positive than in Gram-negative bacteria. This might be due to the fact that Gram-negative bacteria possess an additional structure (cell wall) that shows hydrophilic properties, which might block the pass of the hydrophobic structure of the terpenes [87].

4. Use of Bioactive Compounds from Agro-Industrial Waste in Aquaculture

Aquaculture is an economic sector that shows a broad growth. According to data from the Food and Agriculture Organization of the United Nations [88], aquaculture contributes around half of the production of fish destined for human consumption. Due to the accelerated growth and high demand for aquaculture products, farming has intensified; that is, a greater number of organisms are produced in smaller spaces. Furthermore, other farming factors, such as poor diet, poor water quality, and changes in temperature and pH, might cause stress, suppress the immune system of organisms, and negatively affect their health condition. These conditions might increase the apparition and rapid spread of infectious diseases, which are a major problem for the aquaculture industry due to the economic losses. Traditionally, antibiotics are used to mitigate this problem; however, their unselective use has turned out to be a dangerous solution, due to the appearance of antibiotic-resistant bacteria, as well as the use of these chemicals is undesirable for the final consumer.

Therefore, there is a need to seek alternative options to reduce the problems of disease occurrences by increasing the antioxidant and immune responses in organisms. Recently, there has been a special interest in bioactive compounds, since they have been shown to have multiple properties, such as to promote growth and improve the health of aquatic organisms by reducing oxidative stress and stimulating the immune system, which provides resistance to diseases (Figure 3).

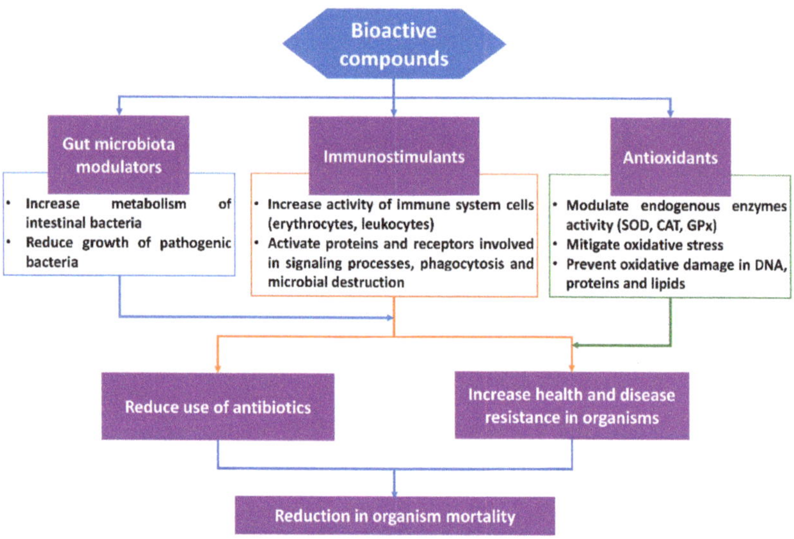

Figure 3. Biological properties demonstrated for bioactive compounds used as food additives for aquatic organisms.

4.1. Bioactive Compounds as Antioxidants in Aquaculture

In aquaculture, there is enough information on the use of bioactive compounds from medicinal plants, as food additives, to increase the antioxidant response and counteract the effects of oxidative stress (lipid oxidation and loss of nutritional quality). In this regard, bioactive compounds from plant residues have been poorly explored.

Corn, rice, wheat, and sorghum are among the most consumed cereals in the world. From the cereal processing a significant amount of residue is generated. This waste could be used for the development of functional foods with antioxidant activity. In this context, some studies have been carried out to evaluate the efficacy of corn and sorghum residues as antioxidant additives in fish feed. Catap et al. [89] reported that the dietary administration of corn silk extract (*Zea mays*) lowered the level of lipid peroxidation in the liver of Nile tilapia (*Oreochromis niloticus*) under paracetamol-induced oxidative stress. Corn silk is an important source of flavonoids such as luteolin, formononetin, maizine, and apigenin [90]. In general, these compounds are recognized for neutralizing reactive oxygen species (ROS) and modulating antioxidant enzyme activities [91]. Therefore, this study might suggest that flavonoids improve the antioxidant response in Nile tilapia liver under induced stress. On the other hand, Lee et al. [92] indicated that the dietary inclusion of residues from the distillation of sorghum (200 g/kg), increases the antioxidant activity and delays the oxidation process of low-density lipoproteins in the plasma of the Lysa (*Mugil cephalus*). Among the main bioactive compounds present in sorghum are phenolic acids (caffeic, ferulic and chlorogenic acids) and flavonoids (apigeninidin, luteolinidin, and naringenin), which have been directly related to its antioxidant activity [93]. Therefore, the efficacy of sorghum to increase antioxidant activity and delay the oxidation of low-density lipoproteins in the plasma of *M. cephalus* is attributed to the ability of PCs to neutralize free radicals.

Waste from fruit processing has received little attention in the aquaculture area, even though it is known to be one of the main sources of bioactive compounds. There are currently few studies related to the use of fruit residues. For example, Giri et al. [94] evaluated the effect of the dietary inclusion of banana peel (*Musa acuminata*) (10, 30, 50 and 70 g/kg) at different feeding times (30 and 60 days) on the formation of MDA and the activity of SOD, GPx, and CAT in rohu (*Labeo rohita*) liver, infected with *Aeromonas hydrophila*. Fish feed including banana peel (50 g/kg and 70 g/kg) showed a significant decrease in MDA levels in both feeding times. Superoxide dismutase and CAT activity increased in fish fed 50 g/kg of banana peel during 60 days of feeding, while GPx activity showed an increase after 30 days of feeding fish the banana peel (30, 50, and 70 g/kg). The authors suggested that the diversity of bioactive compounds identified in the banana peel, such as phenolic acids, flavonoids, and carotenoids, is responsible for improving the hepatic antioxidant response in rohu. Furthermore, Vicente et al. [95] evaluated the effect of orange peel fragment (OPF), as a food additive, on the antioxidant enzyme activity of Nile tilapia subjected to heat/dissolved oxygen-induced stress. In the study, fish were fed diets with different inclusion of OPF (0, 0.2, 0.4, 0.6, and 0.8%) for 70 days. At the end of the feeding trial, fish were subjected to stress conditions (32 °C/2.3 mg/L dissolved oxygen) for three days. Before stress, SOD, CAT, and GPx activities were higher in the non-supplemented group. Nevertheless, after stress, OPF supplementation increased SOD, CAT, and GPx activities. The increase in the antioxidant enzyme activities in Nile tilapia liver could be associated with the presence of hesperidin, a flavonoid, in the orange peel fragments [96]. This flavonoid upregulates the Nrf2 gene expression which improves antioxidant enzyme activity, therefore minimizing oxidative stress [97]. Additionally, Lizárraga-Velázquez et al. [98] reported that supplemented diets with 50 mg and 100 mg of PCs from mango peel extract (MPE) per kg feed decreased MDA levels as a measure of the lipid peroxidation in zebrafish (*Danio rerio*) muscle. Moreover, the authors indicated that the dietary administration of 150 mg and 200 mg PCs from MPE per kg feed increased the hepatic CAT activity without affecting the growth and feed utilization of zebrafish. Antioxidant effects were attributed to the presence of gallic, 2-hydroxycinnamic and protocatechuic acids, mangiferin, quercetin, methyl gallate, and ethyl gallate in the MPE. These latest studies merit further investigation to validate the use of phenolic compounds as feed supplements.

Beta-glucans from mushroom (*Pleurotus pulmonarius*) stalk waste (MSW) have also been explored as antioxidants in aquaculture. For instance, Ahmed et al. [99] evaluated the use of hot water extracts (HWE) from MSW as an additive in fish feed and determined the effect on growth performance and the in vivo antioxidant status of Nile tilapia. When the HWE from MSW, rich in β-glucan content (20.05 ± 0.44%), was added to the diet (10 g/kg), SOD and CAT activities in the liver and kidney were enhanced. The authors mentioned that the effect of β-glucan on SOD and CAT activity might help to prevent the deleterious effects of ROS on organisms. Furthermore, the protection of β-glucans present in HWE of MSW against oxidative stress caused by pH fluctuations in Nile tilapia was also evaluated [100]. Administration of 5 g/kg and 10 g/kg under pre-stress conditions increased SOD and CAT activities in the liver and kidney respectively. Nevertheless, the activity of these enzymes was reduced in liver and kidney samples due to pH changes (5.5 and 10.5). The authors concluded that the supplementation of β-glucans from MSW in the diet for tilapia, enhanced the antioxidant enzyme activities in vivo, which led fish to reduce stress for pH fluctuations and therefore show a normal growth.

4.2. Bioactive Compounds as Modulators of the Immune System and Resistance to Infections

The use of immunostimulants to control aquaculture diseases emerges as an alternative to the use of antibiotics. This topic is gaining particular interest in the scientific community and different sources of obtaining these bioactive have been proposed, such as medicinal plants [101]. However, the use of agro-industrial waste has been less explored, and therefore, the literature in this regard is scarce.

Phenolic compounds are a group of phytochemicals that have been studied in aquaculture as food additives because of their potential as immunostimulants. Particular interest has been taken in PCs from grape seed, a material that is discarded from wine processing. In this regard, Magrone, Fontana, Laforgia, Dragone, Jirillo and Passantino [75] evaluated the effect of grape seed extracts (Canosina Nero di Troia *Vitis vinifera*) on the immune response of juveniles of *Dicentrarchus labrax* L. In this study, the'authors demonstrated that the incorporation of 0.1 and 0.2 g/kg of phenolic extract in feed for *D. labrax*, reduces the levels of IL-1β and IL-6 in the intestine, while the concentration of IFN-γ in the spleen increased. The effect of PCs on the levels of cytokines could be due to the modulation that they exert on the NF-κB factor, which regulates the expression of several cytokines [102]. In addition, the number of MMCs increased. These results show that the diet with polyphenols reduces intestinal inflammation by reducing the levels of proinflammatory cytokines, while the increase in interferon expresses a more robust adaptive immune response. Another aspect is that the increase in the number of MMCs, which contain melanin, is associated with protective functions against pathogens [72]. Furthermore, Arciuli et al. [103] evaluated the effect of PCs extracted from grape seeds on the activity of MMCs, dopa-oxidase, and peroxidase in commercial size fish of *D. labrax*. The administration of 0.2 g/kg in the diet of *D. labrax* increased the activity of dopa-oxidase and peroxidase. These enzymes participate in the synthesis of melanin. The presence of the latter in fish is associated with protective functions against pathological or stress conditions [72]. From these studies, it can be concluded that the addition of PCs from grape seeds increases melanin levels in fish. As previously mentioned, the presence of this pigment is associated with the resistance of organisms against pathogens. Hence PCs could be an option to improve the health status of fish. However, more research is required to validate this effect through a challenge of organisms fed bioactive compounds in the presence of pathogens of interest.

Hoseinifar et al. [104] indicated that the dietary administration of olive waste cake (OWC) (0.5, 2.5, and 5.0 g/kg of feed), increased weight gain (WG) and specific growth rate (SGR), and decreased the feed conversion ratio (FCR) in rainbow trout (*Oncorhynchus mykiss*). The authors also reported that the dietary inclusion of 2.5 g/kg and 5.0 g/kg OWC feed increased total Ig concentration and mucosal lysozyme activity, also up-regulated relative expression of gut IL-8 gene. While, the supplementation of 2.5 g/kg OWC increased the serum lysozyme activity. The immunomodulatory effects of OWC are related to the presence of PCs (hydroxytyrosol, tyrosol, caffeic, *p*-coumaric and vanillic acids, and lutein and lignans) and vitamin E previously identified in olive [105].

Phenolic compounds extracted from olive oil processing waste have also been used in combination with other plant extracts rich in PCs—for instance, chestnut. In this regard, Hoseinifar et al. [106] evaluated the effect of dietary supplementation of a mixture of PCs extracted from olive mill wastewater (OMWW) and chestnut wood (CW) (9:1, OMWW:CW) in concentrations of 0.5, 1.0, and 2.0 g/kg of feed, on the innate immune response of convict cichlid (*Amatitlania nigrofasciata*). The authors reported that mucus total protein levels and lysozyme activities increased in fish fed OMWW:CW. Besides, they indicated that the supplementation of 2.0 g/kg of OMWW:CW increased serum total protein and total Ig levels, as well as peroxidase and radical scavenging activities. The effect of PCs extracted from OMWW:CW (0.5, 0.1, and 2.0 g/kg) on growth performance and innate immune response also have been evaluated in common carp (*C. carpio* L.) [107]. In this study authors reported that supplementation with OMWW:CW increased skin mucus total proteins and Ig levels and lysozyme, peroxidase, and radical scavenging activities in this species. Serum total Ig levels increased in fish fed 0.1 y 0.2 g/kg of OMWW:CW. Besides, PCs from OMWW:CW improved growth (WG and SGR) and feed utilization (FCR) in common carp. In the study, the improve in the immune response in convict cichlid was attributed to PCs such as hydroxytyrosol, tyrosol, and oleuropein, which have been reported in olive, and to tannins identified in CW [108]. However, the mechanism of action by which PCs exert their immunomodulatory effects is still unknown, so efforts should be directed towards that study area. Further studies are necessary with challenges with infectious bacterial or viral diseases to assess the potential of olive waste and CW byproducts as functional feed additives for the aquaculture.

Terpenes, mainly essential oils, have also been studied for their promoter effect of the immune response. Recently it has been reported that the use of essential oils to enhance the immune system response in aquaculture species is a potential alternative to the use of antibiotics [109]. Peels obtained from citrus processing are a good option as a source of essential oil to use them as additives in aquaculture foods [110]. In this context, Acar et al. [111] evaluated the dietary effect of essential oils obtained from orange peel (*Citrus sinensis*) on the growth of Mozambique tilapia (*Oreochromis mossambicus*) and its resistance against the pathogen *Streptococcus iniae*. The fish were fed a control diet, which does not contain essential oils, and three experimental diets (1, 3, and 5 g/kg) for 12 weeks, after which time the fish were challenged by infection with *S. iniae*. Fish fed essential oils, increased lysozyme, and MPO activities. Besides, the addition of essential oils in 1, 3, and 5 g/kg increased fish survival by 48.33%, 46.67%, and 58.33%, respectively. Limonene, a phenolic monoterpene present in the orange peel essential oil, has antibacterial properties and could be responsible for these effects. In general, the results of this study demonstrated that the inclusion of essential oils in diets for tilapia improves the immune response of the fish and therefore may have the potential to be used as antibiotic substitutes. Furthermore, Baba et al. [112] evaluated the effect of essential oils obtained from the lemon peel (*Citrus limon*) on the immune system and resistance against *Edwardsiella tard* in Mozambique tilapia. Fish fed 5 g/kg and 7.5 g/kg of lemon peel essential oil increased the levels of the immuno-hematological parameters, such as the NBT, the number of white cells, hematocrit, and the activity of lysozyme and MPO. After the feeding trial the organisms were subjected to infection by *E. tard*. Fish fed the control diet showed 80% of mortality percentage, while those fed 5, 7.5, and 10 g/kg of essential oils reduced this percentage to 36.6%, 51.6%, and 58.3%, respectively. All the parameters evaluated in this study are important to determine the immune response of the fish. Particularly, lysozyme and MPO have an important role in the elimination of pathogenic microorganisms. Limonene is the main component of the essential oils of lemon peel (54.4%). This compound exerts antimicrobial activity due to its ability to destabilize the bacterial cell membrane [113]. The authors suggest that the addition of 5 g/kg of essential oil of lemon peels into the diet, exerts an immunostimulatory effect, and increases the resistance of Mozambique tilapia against pathogenic bacteria. From these results it can be concluded that citrus peel essential oils are a natural and safe alternative for the formulation of food for aquaculture species.

β-glucans have been widely studied as immunostimulants for organisms of interest in aquaculture, such as shrimp and Nile tilapia [114,115]. β-glucans evaluated are usually obtained from fungi or yeasts [31]. Nevertheless, Chirapongsatonkul et al. [116] recently evaluated the potential of β-glucans obtained from split gill mushroom (*Schizophyllum commune*) cultivation waste as immunostimulant. The authors obtained a crude glucan extract from mycelium containing spent mushroom substrate (SMS) of the *S. commune* to stimulate immune system of Nile tilapia. Fish were injected with 100 μg/mL of glucan extract and after six hours they were challenged with *Aeromonas veronii*. Fish treated with the crude glucan extract showed an increase in immune parameters (Ig, lysozyme) and up-regulation in the expression of cytokine genes (TNF-α, IL-1β, and NF-κB) related to the immune response. Glucan extract treatment also increased the survival rate of Nile tilapia infected with *A. veronii*. These results demonstrate the possibility of use of crude glucan extracts from mushroom cultivation waste to improve the immune response in tilapia.

Cereals, such as oats, barley, and wheat are a rich source of β-glucans [117] and might be used as additives in diets for fish. For instance, Udayangani et al. [118] evaluated the effect of β-glucans from the endosperm of oat grains on the immune response of zebrafish larvae against *E. tard*. Once the larvae hatched, they were kept in solution for three days with two concentrations of β-glucan (100 and 500 μg/mL). Afterward, the larvae were exposed to the pathogen *E. tard* and the expression of cytokines, lysozyme, and survival percentage were determined. Treatment with 500 μg/mL of β-glucan significantly increased the up-regulation of the expression of lysozyme and cytokine genes (TNF-α, IL-1β, IL-10, and IL-12) related to the immune response, as well as the survival rate. Therefore, β-glucans have potential in the aquaculture industry as promoters of the immune system for larval stages of fish. However, waste from cereal processing has not yet been exploited to obtain β-glucans for the development of aquaculture feeds with immunostimulant effects.

4.3. Bioactive Compounds as Modulators of the Intestinal Microbiota

The modulation of the intestinal microbiota through the use of prebiotics (inulin, galactooligosaccharides, and xylooligosaccharides) has received significant attention due to the growing need to (i) replace probiotics due to their high purchasing value, (ii) reduce the incidence of infectious diseases, (iii) improve the health status of aquaculture organisms, and (iv) increase crop production and profitability [78,119].

Inulin is the most widely used prebiotic in aquaculture [78]. The use of this prebiotic increases the lactic acid bacteria population in the gut of surubies (*Pseudoplatystoma* sp.) and beluga sturgeon (*Spindle spindle*) and decreases *Vibrio* spp. in turbot (*Psseta maxima*). These effects are related to inulin inclusion doses [120–122], so further research is still required in this context.

Although there are few studies on the use of prebiotics obtained from agro-industrial wastes, some research has been conducted on prebiotics extracted from cereals, such as wheat. In this regard, Geraylou et al. [123] evaluated the effect of dietary inclusion of wheat bran arabinoxylans (20 g/kg and 40 g/kg), on the composition of the intestinal microbiota of Siberian sturgeon (*Acipenser baerii*). The authors reported that fish fed 20 g/kg and 40 g/kg of arabinoxylans showed an increase in the relative abundance of Eubacteriaceae, Clostridiaceae, Streptococcaceae, and Lactobacillaceae and in Bacillaceae, respectively. Besides, dietary inclusion of wheat bran arabinoxylan oligosaccharides (20 g/kg) modulated the growth of *Lactococcus* sp., *Lactobacillus* sp., *E. budayi*, and several species of the genus *Clostridium*. Furthermore, it has been reported that arabinoxylan oligosaccharides form wheat bran suppress the growth of *Aeromonas* sp., *Citrobacter freundii*, and *Escherichia coli* and increase the content of short-chain fatty acids (acetate and butyrate) in the intestine of Siberian sturgeon [124]. In both studies, the increase in the abundance of the mentioned bacteria is because they possess enzymes (endo-1,4-β-xylanases, α-L-arabinofuranosidases, β-xylosidases, α-glucuronidases, and feruloyl esterases) with capacity to ferment the prebiotics evaluated [125]. It is concluded that wheat bran prebiotics have an impact on the composition of the intestinal microbiota and that the increase in the abundance of lactic acid bacteria and short-chain fatty acids could provide health benefits of Siberian sturgeon.

On the other hand, the use of PCs as prebiotics is of recent interest, so aquaculture studies are scarce. In this context, the effect of PCs of the OMWW on the gut microbiota of narrow clawed crayfish (*Astacus leptodactylus*) was evaluated [126]. Supplementation of 0.5 and 5 g/kg significantly reduced the total intestinal microbiota, with the exclusion of anaerobes and yeasts. This might be because some bacterial groups use PCs as susbtrate, such as lactobacilli. Several PCs present in OMWW, such as, hydroxytyrosol and tyrosol exert antimicrobial properties against bacteria responsible for intestinal infections [127]. The lack of information on the use of PCs extracted from plant residues is a field worth exploring, especially since it is currently known that PCs provide beneficial effects on human health through the increase of bacterial populations beneficial and short-chain fatty acid content [79].

A summary of recent studies in which bioactive compounds obtained from agro-industrial waste were used as feed additives or vaccines, as well as their in vivo effect on antioxidant status, immune system, and microbiota is shown in Table 1.

Table 1. Bioactive compounds from agro-industrial waste and biological effect on aquatic organisms.

Bioactive Compound.	Source (Doses)	Species	Biological Properties	Effect
Phenolic compounds	Corn silk extract (3.5 g/kg)	*Oreochromis niloticus*	Antioxidant, Immunostimulant	↓MDA ↑Lysozyme, against *Aeromonas hydrophila* [89]
Phenolic compounds	Sorghum distillery residue (200 g/kg)	*Mugil cephalus*	Antioxidant	↓LDL oxidation ↑TAC [92]
Phenolic compounds	Banana peel (10, 30, 50, 70 g/kg)	*Labeo rohita*	Antioxidant, Immunostimulant	↓MDA ↑SOD, CAT, GPx ↑IL-1 β, TNF- α ↑Survival rate against *Aeromonas hydrophila* [94]
Phenolic compounds	Orange peel (2, 4, 6, 8 g/kg)	*O. Oreochromis niloticus*	Antioxidant	↑SOD, CAT, GPx [95]
Phenolic compounds: gallic, 2-hydroxycinnamic and protocatechuic acids, quercetin, mangiferine, methyl gallate, ethyl gallate	Mango peel extract (50, 100, 150, 200 mg/kg)	*Danio rerio*	Antioxidant	↑CAT ↓MDA [98]
β-glucans	Mushroom stalk waste extract (5, 10 g/kg)	*Oreochromis niloticus*	Antioxidant	↑SOD, CAT [99,100]
Phenolic compounds: proanthocyanidin, catechins, epicatechins	Grape seeds (0.1, 0.2 g/kg)	*Dicentrarchus labrax* L.	Immunostimulant	↓IL-1β, IL-6. ↑IFN-γ ↑MMCs [75]
Phenolic compounds: catechins, epigallocatechins	Grape seeds (0.1, 0.2 g/kg)	*Dicentrarchus labrax* L.	Immunostimulant	↑Peroxidase ↑Dopa-oxidase ↑MMCs [103]
Phenolic compounds	Olive waste cake (0.5, 2.5, 5 g/kg)	*Oncorhynchus mykiss*	Antioxidant, Immunostimulant	↑SOD, GPx ↑Lysozyme, Ig ↑IL-8 ↓TGF-β [104]

Table 1. *Cont.*

Bioactive Compound.	Source (Doses)	Species	Biological Properties	Effect
Phenolic compounds	Mixture of chestnut wood and olive mill wastewater extract (0.5, 1, 2 g/kg)	*Amatitlania nigrofasciata*	Antioxidant, Immunostimulant	↑Growth performance ↑Ig ↑Lysozyme, peroxidase ↑CAT [106]
Phenolic compounds	Mixture of chestnut wood and olive mill wastewater extract (0.5, 1, 2 g/kg)	*Cyprinus carpio* L.	Immunostimulant	↑Growth performance ↑Ig, lysozyme ↑CAT, peroxidase [107]
Essential oils	Orange peel (1, 3, 5 g/kg)	*Oreochromis mossambicus*	Immunostimulant	↑Lysozyme, MPO ↑Survival rate against *Streptococcus iniae* [111]
Essential oils	Lemon peel (5, 7.5, 10 g/kg)	*Oreochromis mossambicus*	Immunostimulant	↑NBT ↑Lysozyme, MPO ↓Mortality against *Edwardsiella tarda* [112]
Essential oils	Lemon peel (10, 20, 50, 80 g/kg)	*Labeo victorianus*	Immunostimulant	↑Red blood cells, leucocytes, hematocrits, neutrophils ↑Ig, lysozyme ↓Mortality against *Aeromonas hydrophila* [128]
Glucans	Split gill mushroom cultivation waste extract (100 µg/mL)	*Oreochromis niloticus*	Immunostimulant	↑Ig, lysozyme ↑TNF-α, IL-1β, NF-κB ↑Survival rate against *Aeronomas veronii* [116]
Phenolic compounds	Olive mill waste water (0.5, 5 g/kg)	*Astacus leptodactylus*	Antioxidant, Immunostimulant, Microbiota modulation	↑Growth performance ↑CAT, GR ↑Haemocytes ↓Total intestinal bacteria [126]
Arabinoxylans oligosaccharides	Wheat bran (20, 40 g/kg)	*Acipenser baerii*	Microbiota modulation	↑Eubacteriaceae, Clostridiaceae, Streptococcaceae, Lactobacillaceae, Bacillaceae [123]
Arabinoxylans oligosaccharides	Wheat bran (20 g/kg)	*Acipenser baerii*	Immunostimulant, Microbiota modulation	↑Peroxidase, phagocytic activity ↓*Aeromonas* sp., *Citrobacter freundii*, *Escherichia coli* ↑Short-chain fatty acids [124]

CAT—catalase; GPx—glutathione peroxidase; GR—glutathione reductase; IFN-γ—interferon-gamma; Ig—immunoglobulins; IL—interleukin; LDL—low-density lipoproteins; MDA—malondialdehyde; MMCs—melanomacrophage centers; MPO—myeloperoxidase; NBT—nitroblue tetrazolium; NF-κB—nuclear factor kappa light-chain-enhancer of activated B cells; SOD—superoxide dismutase; TAC—total antioxidant capacity; TGF-β—transforming growth factor-beta; TNF-α—tumor necrosis factor-alpha.

5. Conclusions

There is little research aimed to the valorization of agro-industrial waste and its use as a source of bioactive compounds to incorporate them into aquaculture food. Above all, to characterize these wastes in their nutritional aspects, as well as in the quantity and type of bioactive that they present, it is of utmost importance. The activity of bioactive compounds, such as phenolic compounds, terpenes and β-glucans, depends on their chemical structure, the source, the doses, and if it is isolated or in presence of other compounds, as well as the species used as study model. Therefore, these compounds are required to be evaluated in different aquatic organisms of commercial interest, such as shrimp, tilapia, white snook, and snapper, among others, to determine their biological effect, whether antioxidant, immunostimulant, or microbiota modulator. The above reveals the great window of opportunity that exists to explore this topic.

Author Contributions: Writing—original draft preparation, N.L.-L. and C.E.L.-V.; Investigation, N.L.-L., C.E.L.-V., and E.Y.S.-G.; Writing—review and editing, C.H. All authors have read and agreed to the published version of the manuscript.

Funding: This research received no external funding.

Acknowledgments: This manuscript is part of the activities of the Cátedras-CONACYT Project #729: Applied biotechnologies for the development of functional foods for aquaculture (Biotecnologías aplicadas para el desarrollo de alimentos funcionales para acuacultura).

Conflicts of Interest: The authors declare no conflict of interest.

References

1. Nasrin, T.A.A.; Matin, M.A. Valorization of vegetable wastes. In *Food Processing By-Products and Their Utilization*; Anal, A.K., Ed.; John Wiley & Sons Ltd.: Hoboken, NJ, USA, 2017; pp. 53–88.
2. Guaadaoui, A.; Benaicha, S.; Elmajdoub, N.; Bellaoui, M.; Hamal, A. What is a bioactive compound? A combined definition for a preliminary consensus. *Int. J. Nutr. Food Sci.* **2014**, *3*, 174–179. [CrossRef]
3. Sagar, N.A.; Pareek, S.; Sharma, S.; Yahia, E.M.; Lobo, M.G. Fruit and vegetable waste: Bioactive compounds, their extraction, and possible utilization. *Compr. Rev. Food Sci. Food Saf.* **2018**, *17*, 512–531. [CrossRef]
4. González-Sánchez, M.E.; Pérez-Fabiel, S.; Wong-Villarreal, A.; Bello-Mendoza, R.; Yañez-Ocampo, G. Residuos agroindustriales con potencial para la producción de metano mediante la digestión anaerobia. *Rev. Argent. Microbiol.* **2015**, *47*, 229–235. [CrossRef]
5. Ayala-Zavala, J.F.; Vega-Vega, V.; Rosas-Domínguez, C.; Palafox-Carlos, H.; Villa-Rodriguez, J.A.; Siddiqui, M.W.; Dávila-Aviña, J.E.; González-Aguilar, G.A. Agro-industrial potential of exotic fruit byproducts as a source of food additives. *Food Res. Int.* **2011**, *44*, 1866–1874. [CrossRef]
6. Gemede, H.F.; Ratta, N. Antinutritional factors in plant foods: Potential health benefits and adverse effects. *Int. J. Nutr. Food Sci.* **2014**, *3*, 284–289. [CrossRef]
7. Nour-Eldin, H.H.; Madsen, S.R.; Engelen, S.; Jorgensen, M.E.; Olsen, C.E.; Andersen, J.S.; Seynnaeve, D.; Verhoye, T.; Fulawka, R.; Denolf, P.; et al. Reduction of antinutritional glucosinolates in *Brassica* oilseeds by mutation of genes encoding transporters. *Nat. Biotechnol.* **2017**, *35*, 377–382. [CrossRef] [PubMed]
8. Veneziani, G.; Novelli, E.; Esposto, S.; Taticchi, A.; Servili, M. Chapter 11—Applications of recovered bioactive compounds in food products. In *Olive Mill Waste*; Galanakis, C.M., Ed.; Academic Press: Cambridge, MA, USA, 2017; pp. 231–253.
9. Vermerris, W.; Nicholson, R. Families of phenolic compounds and means of classification. In *Phenolic Compound Biochemistry*; Springer: Dordrecht, The Netherlands, 2006; pp. 1–34.
10. Ferreira, I.C.F.R.; Martins, N.; Barros, L. Chapter One—Phenolic compounds and its bioavailability: In vitro bioactive compounds or health promoters? In *Advances in Food and Nutrition Research*; Toldrá, F., Ed.; Academic Press: Cambridge, MA, USA, 2017; Volume 82, pp. 1–44.
11. Martins, N.; Barros, L.; Ferreira, I.C.F.R. In vivo antioxidant activity of phenolic compounds: Facts and gaps. *Trends Food Sci. Technol.* **2016**, *48*, 1–12. [CrossRef]

12. Soidrou, S.H.; Bousta, D.; Lachkar, M.; Hassane, S.O.S.; Youbi-Hamsas, A.E.; Mansouri, L.E.; Benjilali, J.; El-Hajaji, H.; Farah, A. *Immunomodulatory Activity of Phenolic Fraction from Piper Borbonense and Cassytha Filiformis Growing in Comoros Islands*; Springer: Dordrecht, The Netherlands, 2014; pp. 105–112.
13. Espín, J.C.; González-Sarrías, A.; Tomás-Barberán, F.A. The gut microbiota: A key factor in the therapeutic effects of (poly)phenols. *Biochem. Pharmacol.* **2017**, *139*, 82–93. [CrossRef]
14. Lu, J.; Fu, X.; Liu, T.; Zheng, Y.; Chen, J.; Luo, F. Phenolic composition, antioxidant, antibacterial and anti-inflammatory activities of leaf and stem extracts from *Cryptotaenia japonica* Hassk. *Ind. Crop. Prod.* **2018**, *122*, 522–532. [CrossRef]
15. Morais-Braga, M.F.B.; Souza, T.M.; Santos, K.K.A.; Guedes, G.M.M.; Andrade, J.C.; Vega, C.; Rolón, M.; Costa, J.G.M.; Saraiva, A.A.F.; Coutinho, H.D.M. Phenol composition, cytotoxic and anti-kinetoplastidae activities of *Lygodium venustum* SW. (Lygodiaceae). *Exp. Parasitol.* **2013**, *134*, 178–182. [CrossRef]
16. Zhang, X.-L.; Guo, Y.-S.; Wang, C.-H.; Li, G.-Q.; Xu, J.-J.; Chung, H.Y.; Ye, W.-C.; Li, Y.-L.; Wang, G.-C. Phenolic compounds from *Origanum vulgare* and their antioxidant and antiviral activities. *Food Chem.* **2014**, *152*, 300–306. [CrossRef] [PubMed]
17. Gabr, S.; Nikles, S.; Pferschy Wenzig, E.M.; Ardjomand-Woelkart, K.; Hathout, R.M.; El-Ahmady, S.; Motaal, A.A.; Singab, A.; Bauer, R. Characterization and optimization of phenolics extracts from *Acacia* species in relevance to their anti-inflammatory activity. *Biochem. Syst. Ecol.* **2018**, *78*, 21–30. [CrossRef]
18. Kumar, S.S.; Patra, A. Evaluation of phenolic composition, antioxidant, anti-inflammatory and anticancer activities of *Polygonatum verticillatum* (L.). *J. Integr. Med.* **2018**, *16*, 273–282. [CrossRef] [PubMed]
19. Yousefian, M.; Shakour, N.; Hosseinzadeh, H.; Hayes, A.W.; Hadizadeh, F.; Karimi, G. The natural phenolic compounds as modulators of NADPH oxidases in hypertension. *Phytomedicine* **2018**, *55*, 200–213. [CrossRef] [PubMed]
20. Leicach, S.R.; Chludil, H.D. Chapter 9—Plant secondary metabolites: Structure–activity relationships in human health prevention and treatment of common diseases. In *Studies in Natural Products Chemistry*; Attaur, R., Ed.; Elsevier: Amsterdam, The Netherlands, 2014; Volume 42, pp. 267–304.
21. Fraga, C.G.; Galleano, M.; Verstraeten, S.V.; Oteiza, P.I. Basic biochemical mechanisms behind the health benefits of polyphenols. *Mol. Asp. Med.* **2010**, *31*, 435–445. [CrossRef]
22. Tarahovsky, Y.S.; Kim, Y.A.; Yagolnik, E.A.; Muzafarov, E.N. Flavonoid–membrane interactions: Involvement of flavonoid–metal complexes in raft signaling. *Biochim. Biophys. Acta* **2014**, *1838*, 1235–1246. [CrossRef]
23. Hussain, T.; Tan, B.; Liu, G.; Murtaza, G.; Rahu, N.; Saleem, M.; Yin, Y. Modulatory mechanism of polyphenols and Nrf2 signaling pathway in LPS challenged pregnancy disorders. *Oxid. Med. Cell. Longev.* **2017**, *2017*, 8254289. [CrossRef]
24. Kirby, J.; Keasling, J.D. Biosynthesis of plant isoprenoids: Perspectives for microbial engineering. *Annu. Rev. Plant Biol.* **2009**, *60*, 335–355. [CrossRef]
25. Babahmad, R.A.; Aghraz, A.; Boutafda, A.; Papazoglou, E.G.; Tarantilis, P.A.; Kanakis, C.; Hafidi, M.; Ouhdouch, Y.; Outzourhit, A.; Ouhammou, A. Chemical composition of essential oil of *Jatropha curcas* L. leaves and its antioxidant and antimicrobial activities. *Ind. Crop. Prod.* **2018**, *121*, 405–410. [CrossRef]
26. Hassoun, A.; Emir Çoban, Ö. Essential oils for antimicrobial and antioxidant applications in fish and other seafood products. *Trends Food Sci. Technol.* **2017**, *68*, 26–36. [CrossRef]
27. Amorati, R.; Foti, M.C.; Valgimigli, L. Antioxidant Activity of Essential Oils. *J. Agric. Food Chem.* **2013**, *61*, 10835–10847. [CrossRef] [PubMed]
28. Hyldgaard, M.; Mygind, T.; Meyer, R.L. Essential oils in food preservation: Mode of action, synergies, and interactions with food matrix components. *Front. Microbiol.* **2012**, *3*, 12. [CrossRef] [PubMed]
29. Makki, K.; Deehan, E.C.; Walter, J.; Bäckhed, F. The impact of dietary fiber on gut microbiota in host health and disease. *Cell Host Microbe* **2018**, *23*, 705–715. [CrossRef]
30. Deehan, E.C.; Duar, R.M.; Armet, A.M.; Perez-Muñoz, M.E.; Jin, M.; Walter, J. Modulation of the gastrointestinal microbiome with nondigestible fermentable carbohydrates to improve human health. *Microbiol. Spectr.* **2017**, *5*. [CrossRef]
31. Medina-Gali, R.M.; Ortega-Villaizan, M.D.M.; Mercado, L.; Novoa, B.; Coll, J.; Perez, L. Beta-glucan enhances the response to SVCV infection in zebrafish. *Dev. Comp. Immunol.* **2018**, *84*, 307–314. [CrossRef]
32. Gidley, M.J.; Nishinari, K. Chapter 2.2—Physico-chemistry of (1,3)-β-Glucans. In *Chemistry, Biochemistry, and Biology of 1-3 Beta Glucans and Related Polysaccharides*; Bacic, A., Fincher, G.B., Stone, B.A., Eds.; Academic Press: San Diego, CA, USA, 2009; pp. 47–118.

33. Stone, B.A. Chapter 2.1—Chemistry of β-Glucans. In *Chemistry, Biochemistry, and Biology of 1-3 Beta Glucans and Related Polysaccharides*; Bacic, A., Fincher, G.B., Stone, B.A., Eds.; Academic Press: San Diego, CA, USA, 2009; pp. 5–46.
34. Jin, Y.; Li, P.; Wang, F. β-glucans as potential immunoadjuvants: A review on the adjuvanticity, structure-activity relationship and receptor recognition properties. *Vaccine* **2018**, *36*, 5235–5244. [CrossRef]
35. Bischoff, K.L. Chapter 40—Glucosinolates. In *Nutraceuticals*; Gupta, R.C., Ed.; Academic Press: Boston, MA, USA, 2016; pp. 551–554.
36. Goncharov, N.; Orekhov, A.N.; Voitenko, N.; Ukolov, A.; Jenkins, R.; Avdonin, P. Chapter 41—Organosulfur compounds as nutraceuticals. In *Nutraceuticals*; Gupta, R.C., Ed.; Academic Press: Boston, MA, USA, 2016; pp. 555–568.
37. Yalcin, H.; Çapar, T.D. Bioactive compounds of fruits and vegetables. In *Minimally Processed Refrigerated Fruits and Vegetables*; Yildiz, F., Wiley, R.C., Eds.; Springer: Boston, MA, USA, 2017; pp. 723–745. [CrossRef]
38. Vig, A.P.; Rampal, G.; Thind, T.S.; Arora, S. Bio-protective effects of glucosinolates—A review. *LWT Food Sci. Technol.* **2009**, *42*, 1561–1572. [CrossRef]
39. Kojima, M.; Oawa, K. Studies on the effect of isothyocianates and their analogues on microorganisms. I. Effect of isothiocyanates on the oxygen uptake of yeasts. *J. Ferment. Technol.* **1971**, *49*, 740–746.
40. Zsolnai, T. Antimicrobial effect of thiocyanates and isothiocyanates. *Arztl. Forsch.* **1966**, *16*, 870–876.
41. Yang, Y.; Laval, S.; Yu, B. Chapter 2—Chemical synthesis of saponins. In *Advances in Carbohydrate Chemistry and Biochemistry*; Horton, D., Ed.; Academic Press: Cambridge, MA, USA, 2014; Volume 71, pp. 137–226.
42. Sidana, J.; Singh, B.; Sharma, O.P. Saponins of agave: Chemistry and bioactivity. *Phytochemistry* **2016**, *130*, 22–46. [CrossRef]
43. Singh, B.; Singh, J.P.; Singh, N.; Kaur, A. Saponins in pulses and their health promoting activities: A review. *Food Chem.* **2017**, *233*, 540–549. [CrossRef] [PubMed]
44. Sarikahya, N.B.; Nalbantsoy, A.; Top, H.; Gokturk, R.S.; Sumbul, H.; Kirmizigul, S. Immunomodulatory, hemolytic and cytotoxic activity potentials of triterpenoid saponins from eight *Cephalaria* species. *Phytomedicine* **2018**, *38*, 135–144. [CrossRef] [PubMed]
45. Press, J.B.; Reynolds, R.C.; May, R.D.; Marciani, D.J. Structure/function relationships of immunostimulating saponins. In *Studies in Natural Products Chemistry*; Attaur, R., Ed.; Elsevier: Amsterdam, The Netherlands, 2000; Volume 24, pp. 131–174.
46. Jacob, M.C.; Favre, M.; Bensa, J.-C. Membrane cell permeabilisation with saponin and multiparametric analysis by flow cytometry. *Cytometry* **1991**, *12*, 550–558. [CrossRef] [PubMed]
47. Arabski, M.; Wegierek-Ciuk, A.; Czerwonka, G.; Lankoff, A.; Kaca, W. Effects of saponins against clinical *E. coli* strains and eukaryotic cell line. *J. Biomed. Biotechnol.* **2012**, *2012*, 6. [CrossRef]
48. Francis, G.; Makkar, H.P.; Becker, K. Antinutritional factors present in plant-derived alternate fish feed ingredients and their effects in fish. *Aquaculture* **2001**, *193*, 197–227. [CrossRef]
49. Omnes, M.-H.; Le Goasduff, J.; Le Delliou, H.; Le Bayon, N.; Quazuguel, P.; Robin, J.H. Effects of dietary tannin on growth, feed utilization and digestibility, and carcass composition in juvenile European seabass (*Dicentrarchus labrax* L.). *Aquacult. Rep.* **2017**, *6*, 21–27. [CrossRef]
50. Fernandes, R.D.P.P.; Trindade, M.A.; de Melo, M.P. Chapter 2—Natural antioxidants and food applications: Healthy perspectives. In *Alternative and Replacement Foods*; Holban, A.M., Grumezescu, A.M., Eds.; Academic Press: Cambridge, MA, USA, 2018; pp. 31–64.
51. Ayala, A.; Muñoz, M.F.; Argüelles, S. Lipid peroxidation: Production, metabolism, and signaling mechanisms of malondialdehyde and 4-Hydroxy-2-Nonenal. *Oxid. Med. Cell. Longev.* **2014**, *2014*, 360438. [CrossRef]
52. Pizzino, G.; Irrera, N.; Cucinotta, M.; Pallio, G.; Mannino, F.; Arcoraci, V.; Squadrito, F.; Altavilla, D.; Bitto, A. Oxidative stress: Harms and benefits for human health. *Oxid. Med. Cell. Longev.* **2017**, *2017*, 8416763. [CrossRef]
53. Pillon, N.J.; Croze, M.L.; Vella, R.E.; Soulère, L.; Lagarde, M.; Soulage, C.O. The lipid peroxidation by-product 4-hydroxy-2-nonenal (4-HNE) induces insulin resistance in skeletal muscle through both carbonyl and oxidative stress. *Endocrinology* **2012**, *153*, 2099–2111. [CrossRef]
54. Kumar, S.; Krishna, C.R.; Preedy, V.R. Chapter 20—Assessment of antioxidant potential of dietary components. In *HIV/AIDS*; Preedy, V.R., Watson, R.R., Eds.; Academic Press: Cambridge, MA, USA, 2018; pp. 239–253.
55. Kim, K.M.; Ki, S.H. Chapter 28—Nrf2: A key regulator of redox signaling in liver diseases. In *Liver Pathophysiology*; Muriel, P., Ed.; Academic Press: Boston, MA, USA, 2017; pp. 355–374.

56. Lingappan, K. NF-κB in oxidative stress. *Curr. Opin. Toxicol.* **2018**, *7*, 81–86. [CrossRef]
57. Zhou, Y.; Jiang, Z.; Lu, H.; Xu, Z.; Tong, R.; Shi, J.; Jia, G. Recent advances of natural polyphenols activators for Keap1-Nrf2 signaling pathway. *Chem. Biodivers.* **2019**, *16*, e1900400. [CrossRef] [PubMed]
58. Ishida, Y.; Ohta, K.; Naruse, T.; Kato, H.; Fukui, A.; Shigeishi, H.; Nishi, H.; Tobiume, K.; Takechi, M. Candida albicans β-glucan-containing particles increase HO-1 expression in oral keratinocytes via a reactive oxygen species/p38 mitogen-activated protein kinase/Nrf2 pathway. *Infect. Immun.* **2018**, *86*, e00575-17. [CrossRef] [PubMed]
59. Eggler, A.L.; Savinov, S.N. Chemical and biological mechanisms of phytochemical activation of Nrf2 and importance in disease prevention. *Recent. Adv. Phytochem.* **2013**, *43*, 121–155. [CrossRef]
60. Inoue, Y.; Shimazawa, M.; Nagano, R.; Kuse, Y.; Takahashi, K.; Tsuruma, K.; Hayashi, M.; Ishibashi, T.; Maoka, T.; Hara, H. Astaxanthin analogs, adonixanthin and lycopene, activate Nrf2 to prevent light-induced photoreceptor degeneration. *J. Pharmacol. Sci.* **2017**, *134*, 147–157. [CrossRef]
61. Yahfoufi, N.; Alsadi, N.; Jambi, M.; Matar, C. The immunomodulatory and anti-inflammatory role of polyphenols. *Nutrients* **2018**, *10*, 1618. [CrossRef] [PubMed]
62. Surh, Y.-J.; Na, H.-K. NF-kappaB and Nrf2 as prime molecular targets for chemoprevention and cytoprotection with anti-inflammatory and antioxidant phytochemicals. *Genes Nutr.* **2008**, *2*, 313–317. [CrossRef]
63. Zhao, B.; Ren, B.; Guo, R.; Zhang, W.; Ma, S.; Yao, Y.; Yuan, T.; Liu, Z.; Liu, X. Supplementation of lycopene attenuates oxidative stress induced neuroinflammation and cognitive impairment via Nrf2/NF-κB transcriptional pathway. *Food Chem. Toxicol.* **2017**, *109*, 505–516. [CrossRef]
64. Garcia, Y.J.; Rodríguez-Malaver, A.J.; Peñaloza, N. Lipid peroxidation measurement by thiobarbituric acid assay in rat cerebellar slices. *J. Neurosci. Methods* **2005**, *144*, 127–135. [CrossRef] [PubMed]
65. Ighodaro, O.M.; Akinloye, O.A. First line defence antioxidants-superoxide dismutase (SOD), catalase (CAT) and glutathione peroxidase (GPX): Their fundamental role in the entire antioxidant defence grid. *Alex. J. Med.* **2018**, *54*, 287–293. [CrossRef]
66. Vallejos-Vidal, E.; Reyes-López, F.; Teles, M.; MacKenzie, S. The response of fish to immunostimulant diets. *Fish Shellfish Immunol.* **2016**, *56*, 34–69. [CrossRef]
67. Caipang, C.M.A.; Lazado, C.C. 9—Nutritional impacts on fish mucosa: Immunostimulants, pre- and probiotics. In *Mucosal Health in Aquaculture*; Beck, B.H., Peatman, E., Eds.; Academic Press: San Diego, CA, USA, 2015; pp. 211–272.
68. Syngai, G.G.; Ahmed, G. Chapter 11—Lysozyme: A natural antimicrobial enzyme of interest in food applications. In *Enzymes in Food Biotechnology*; Kuddus, M., Ed.; Academic Press: Cambridge, MA, USA, 2019; pp. 169–179.
69. Aratani, Y. Myeloperoxidase: Its role for host defense, inflammation, and neutrophil function. *Arch. Biochem. Biophys.* **2018**, *640*, 47–52. [CrossRef] [PubMed]
70. Thomas, D.C. The phagocyte respiratory burst: Historical perspectives and recent advances. *Immunol. Lett.* **2017**, *192*, 88–96. [CrossRef] [PubMed]
71. Clos, T.W.D.; Mold, C. 20—Complement and complement deficiencies. In *Clinical Immunology*, 3rd ed.; Rich, R.R., Fleisher, T.A., Shearer, W.T., Schroeder, H.W., Frew, A.J., Weyand, C.M., Eds.; Mosby: Edinburgh, UK, 2008; pp. 305–325.
72. Agius, C.; Roberts, R.J. Melano-macrophage centres and their role in fish pathology. *J. Fish Dis.* **2003**, *26*, 499–509. [CrossRef] [PubMed]
73. Gilmore, T.D. Introduction to NF-κB: Players, pathways, perspectives. *Oncogene* **2006**, *25*, 6680–6684. [CrossRef]
74. Nair, M.P.N.; Kandaswami, C.; Mahajan, S.; Chadha, K.C.; Chawda, R.; Nair, H.; Kumar, N.; Nair, R.E.; Schwartz, S.A. The flavonoid, quercetin, differentially regulates Th-1 (IFNγ) and Th-2 (IL4) cytokine gene expression by normal peripheral blood mononuclear cells. *Biochim. Biophys. Acta* **2002**, *1593*, 29–36. [CrossRef]
75. Magrone, T.; Fontana, S.; Laforgia, F.; Dragone, T.; Jirillo, E.; Passantino, L. Administration of a polyphenol-enriched feed to farmed sea bass (*Dicentrarchus labrax* L.) modulates intestinal and spleen immune responses. *Oxid. Med. Cell. Longev.* **2016**, *2016*, 11. [CrossRef]
76. Chi, G.; Wei, M.; Xie, X.; Soromou, L.W.; Liu, F.; Zhao, S. Suppression of MAPK and NF-κB pathways by limonene contributes to attenuation of lipopolysaccharide-induced inflammatory responses in acute lung injury. *Inflammation* **2013**, *36*, 501–511. [CrossRef]

77. Petit, J.; Bailey, E.C.; Wheeler, R.T.; de Oliveira, C.A.F.; Forlenza, M.; Wiegertjes, G.F. Studies into β-glucan recognition in fish suggests a key role for the C-Type lectin pathway. *Front. Immunol.* **2019**, *10*, 280. [CrossRef]
78. Ringø, E.; Zhou, Z.; Vecino, J.L.G.; Wadsworth, S.; Romero, J.; Krogdahl, Å.; Olsen, R.E.; Dimitroglou, A.; Foey, A.; Davies, S.; et al. Effect of dietary components on the gut microbiota of aquatic animals. A never-ending story? *Aquacult. Nutr.* **2016**, *22*, 219–282. [CrossRef]
79. Tomás-Barberán, F.A.; Selma, M.V.; Espín, J.C. Interactions of gut microbiota with dietary polyphenols and consequences to human health. *Curr. Opin. Clin. Nutr. Metab. Care* **2016**, *19*, 471–476. [CrossRef]
80. Volman, J.J.; Ramakers, J.D.; Plat, J. Dietary modulation of immune function by β-glucans. *Physiol. Behav.* **2008**, *94*, 276–284. [CrossRef] [PubMed]
81. Maheshwari, G.; Sowrirajan, S.; Joseph, B. β-Glucan, a dietary fiber in effective prevention of lifestyle diseases—An insight. *Bioact. Carbohydr. Diet. Fibre* **2019**, *19*, 100187. [CrossRef]
82. Alminger, M.; Eklund-Jonsson, C. Whole-grain cereal products based on a high-fibre barley or oat genotype lower post-prandial glucose and insulin responses in healthy humans. *Eur. J. Nutr.* **2008**, *47*, 294. [CrossRef] [PubMed]
83. Jayachandran, M.; Chen, J.; Chung, S.S.M.; Xu, B. A critical review on the impacts of β-glucans on gut microbiota and human health. *J. Nutr. Biochem.* **2018**, *61*, 101–110. [CrossRef] [PubMed]
84. Duda-Chodak, A.; Tarko, T.; Satora, P.; Sroka, P. Interaction of dietary compounds, especially polyphenols, with the intestinal microbiota: A review. *Eur. J. Nutr.* **2015**, *54*, 325–341. [CrossRef]
85. Chen, Y.; Li, Q.; Zhao, T.; Zhang, Z.; Mao, G.; Feng, W.; Wu, X.; Yang, L. Biotransformation and metabolism of three mulberry anthocyanin monomers by rat gut microflora. *Food Chem.* **2017**, *237*, 887–894. [CrossRef]
86. Sun, H.; Chen, Y.; Cheng, M.; Zhang, X.; Zheng, X.; Zhang, Z. The modulatory effect of polyphenols from green tea, oolong tea and black tea on human intestinal microbiota in vitro. *J. Food Sci. Technol.* **2018**, *55*, 399–407. [CrossRef] [PubMed]
87. Tariq, S.; Wani, S.; Rasool, W.; Shafi, K.; Bhat, M.A.; Prabhakar, A.; Shalla, A.H.; Rather, M.A. A comprehensive review of the antibacterial, antifungal and antiviral potential of essential oils and their chemical constituents against drug-resistant microbial pathogens. *Microb. Pathog.* **2019**, *134*, 103580. [CrossRef]
88. FAO. *The State of World Fisheries and Aquaculture 2018-Meeting the Sustainable Development Goals*; License CC BY-NC-SA 3.0 IGO; Food and Agriculture Organization of the United Nations: Rome, Italy, 2018; p. 210.
89. Catap, E.S.; Jimenez, M.R.R.; Tumbali, M.P.B. Immunostimulatory and anti-oxidative properties of corn silk from *Zea mays* L. in Nile tilapia, *Oreochromis niloticus*. *Int. J. Fish. Aquac.* **2015**, *7*, 30–36. [CrossRef]
90. Žilić, S.; Janković, M.; Basić, Z.; Vančetović, J.; Maksimović, V. Antioxidant activity, phenolic profile, chlorophyll and mineral matter content of corn silk (*Zea mays* L.): Comparison with medicinal herbs. *J. Cereal Sci.* **2016**, *69*, 363–370. [CrossRef]
91. Panche, A.N.; Diwan, A.D.; Chandra, S.R. Flavonoids: An overview. *J. Nutr. Sci.* **2016**, *5*, e47. [CrossRef]
92. Lee, S.M.; Cheng, H.L.; Pan, B.S. LDL oxidation, antioxidant capacity and growth of cultured grey mullet (*Mugil cephalus*) fed dietary sorghum distillery residue pretreated with polyethylene glycol. *J. Agric. Food Chem.* **2009**, *57*, 7877–7882. [CrossRef]
93. Wu, G.; Bennett, S.J.; Bornman, J.F.; Clarke, M.W.; Fang, Z.; Johnson, S.K. Phenolic profile and content of sorghum grains under different irrigation managements. *Food Res. Int.* **2017**, *97*, 347–355. [CrossRef]
94. Giri, S.S.; Jun, J.W.; Sukumaran, V.; Park, S.C. Dietary Administration of banana (*Musa acuminata*) peel flour affects the growth, antioxidant status, cytokine responses, and disease susceptibility of rohu, *Labeo rohita*. *J. Immunol. Res.* **2016**, *2016*, 4086591. [CrossRef]
95. Vicente, I.S.T.; Fleuri, L.F.; Carvalho, P.L.P.F.; Guimarães, M.G.; Naliato, R.F.; Müller, H.d.C.; Sartori, M.M.P.; Pezzato, L.E.; Barros, M.M. Orange peel fragment improves antioxidant capacity and haematological profile of Nile tilapia subjected to heat/dissolved oxygen-induced stress. *Aquacult. Res.* **2019**, *50*, 80–92. [CrossRef]
96. Zou, Z.; Xi, W.; Hu, Y.; Nie, C.; Zhou, Z. Antioxidant activity of Citrus fruits. *Food Chem.* **2016**, *196*, 885–896. [CrossRef]
97. Parhiz, H.; Roohbakhsh, A.; Soltani, F.; Rezaee, R.; Iranshahi, M. Antioxidant and anti-inflammatory properties of the citrus flavonoids hesperidin and hesperetin: An updated review of their molecular mechanisms and experimental models. *Phytother. Res.* **2015**, *29*, 323–331. [CrossRef]
98. Lizárraga-Velázquez, C.E.; Hernández, C.; González-Aguilar, G.A.; Heredia, J.B. Effect of dietary intake of phenolic compounds from mango peel extract on growth, lipid peroxidation and antioxidant enzyme activities in zebrafish (*Danio rerio*). *Lat. Am. J. Aquat. Res.* **2019**, *47*, 602–611. [CrossRef]

99. Ahmed, M.; Abdullah, N.; Yusof, H.M.; Shuib, A.S.; Razak, S.A. Improvement of growth and antioxidant status in Nile tilapia, *Oreochromis niloticus*, fed diets supplemented with mushroom stalk waste hot water extract. *Aquacult. Res.* **2017**, *48*, 1146–1157. [CrossRef]
100. Ahmed, M.; Abdullah, N.; Shuib, A.S.; Razak, S.A. Influence of raw polysaccharide extract from mushroom stalk waste on growth and pH perturbation induced-stress in Nile tilapia, *Oreochromis niloticus*. *Aquaculture* **2017**, *468*, 60–70. [CrossRef]
101. Vaseeharan, B.; Thaya, R. Medicinal plant derivatives as immunostimulants: An alternative to chemotherapeutics and antibiotics in aquaculture. *Aquacult. Int.* **2014**, *22*, 1079–1091. [CrossRef]
102. Romier, B.; Van De Walle, J.; During, A.; Larondelle, Y.; Schneider, Y.-J. Modulation of signalling nuclear factor-κB activation pathway by polyphenols in human intestinal Caco-2 cells. *Br. J. Nutr.* **2008**, *100*, 542–551. [CrossRef] [PubMed]
103. Arciuli, M.; Fiocco, D.; Fontana, S.; Arena, M.P.; Frassanito, M.A.; Gallone, A. Administration of a polyphenol-enriched feed to farmed sea bass (*Dicentrarchus labrax* L.): Kidney melanomacrophages response. *Fish Shellfish Immunol.* **2017**, *68*, 404–410. [CrossRef] [PubMed]
104. Hoseinifar, S.H.; Shakouri, M.; Yousefi, S.; Van Doan, H.; Shafiei, S.; Yousefi, M.; Mazandarani, M.; Torfi Mozanzadeh, M.; Tulino, M.G.; Faggio, C. Humoral and skin mucosal immune parameters, intestinal immune related genes expression and antioxidant defense in rainbow trout (*Oncorhynchus mykiss*) fed olive (*Olea europea* L.) waste. *Fish Shellfish Immunol.* **2020**, *100*, 171–178. [CrossRef] [PubMed]
105. Khalili, F.; Taghian Dinani, S. Extraction of phenolic compounds from olive-waste cake using ultrasonic process. *J. Food Meas. Charact.* **2018**, *12*, 974–981. [CrossRef]
106. Hoseinifar, S.H.; Jahazi, M.A.; Nikdehghan, N.; Van Doan, H.; Volpe, M.G.; Paolucci, M. Effects of dietary polyphenols from agricultural by-products on mucosal and humoral immune and antioxidant responses of convict cichlid (*Amatitlania nigrofasciata*). *Aquaculture* **2020**, *517*, 734790. [CrossRef]
107. Jahazi, M.A.; Hoseinifar, S.H.; Jafari, V.; Hajimoradloo, A.; Van Doan, H.; Paolucci, M. Dietary supplementation of polyphenols positively affects the innate immune response, oxidative status, and growth performance of common carp, *Cyprinus carpio* L. *Aquaculture* **2020**, *517*, 734709. [CrossRef]
108. Ricci, A.; Olejar, K.J.; Parpinello, G.P.; Kilmartin, P.A.; Versari, A. Application of fourier transform infrared (FTIR) spectroscopy in the characterization of tannins. *Appl. Spectrosc. Rev.* **2015**, *50*, 407–442. [CrossRef]
109. Da Cunha, J.A.; de Ávila Scheeren, C.; Fausto, V.P.; de Melo, L.D.W.; Henneman, B.; Frizzo, C.P.; de Almeida Vaucher, R.; Castagna de Vargas, A.; Baldisserotto, B. The antibacterial and physiological effects of pure and nanoencapsulated *Origanum majorana* essential oil on fish infected with *Aeromonas hydrophila*. *Microb. Pathog.* **2018**, *124*, 116–121. [CrossRef]
110. Sharma, K.; Mahato, N.; Cho, M.H.; Lee, Y.R. Converting citrus wastes into value-added products: Economic and environmently friendly approaches. *Nutrition* **2017**, *34*, 29–46. [CrossRef]
111. Acar, Ü.; Kesbiç, O.S.; Yılmaz, S.; Gültepe, N.; Türker, A. Evaluation of the effects of essential oil extracted from sweet orange peel (*Citrus sinensis*) on growth rate of tilapia (*Oreochromis mossambicus*) and possible disease resistance against Streptococcus iniae. *Aquaculture* **2015**, *437*, 282–286. [CrossRef]
112. Baba, E.; Acar, Ü.; Öntaş, C.; Kesbiç, O.S.; Yılmaz, S. Evaluation of *Citrus limon* peels essential oil on growth performance, immune response of Mozambique tilapia *Oreochromis mossambicus* challenged with *Edwardsiella tarda*. *Aquaculture* **2016**, *465*, 13–18. [CrossRef]
113. Bakkali, F.; Averbeck, S.; Averbeck, D.; Idaomar, M. Biological effects of essential oils—A review. *Food Chem. Toxicol.* **2008**, *46*, 446–475. [CrossRef]
114. Salah, A.S.; El Nahas, A.F.; Mahmoud, S. Modulatory effect of different doses of β-1,3/1,6-glucan on the expression of antioxidant, inflammatory, stress and immune-related genes of *Oreochromis niloticus* challenged with *Streptococcus iniae*. *Fish Shellfish Immunol.* **2017**, *70*, 204–213. [CrossRef] [PubMed]
115. Wu, Y.-S.; Liau, S.-Y.; Huang, C.-T.; Nan, F.-H. Beta 1,3/1,6-glucan and vitamin C immunostimulate the non-specific immune response of white shrimp (*Litopenaeus vannamei*). *Fish Shellfish Immunol.* **2016**, *57*, 269–277. [CrossRef] [PubMed]
116. Chirapongsatonkul, N.; Mueangkan, N.; Wattitum, S.; U-taynapun, K. Comparative evaluation of the immune responses and disease resistance of Nile tilapia (*Oreochromis niloticus*) induced by yeast β-glucan and crude glucan derived from mycelium in the spent mushroom substrate of *Schizophyllum commune*. *Aquacult. Rep.* **2019**, *15*, 100205. [CrossRef]

117. Nie, S.; Cui, S.W.; Xie, M. Chapter 9—Cereal beta-glucan. In *Bioactive Polysaccharides*; Nie, S., Cui, S.W., Xie, M., Eds.; Academic Press: Cambridge, MA, USA, 2018; pp. 445–482.
118. Udayangani, R.M.C.; Dananjaya, S.H.S.; Fronte, B.; Kim, C.-H.; Lee, J.; De Zoysa, M. Feeding of nano scale oats β-glucan enhances the host resistance against *Edwardsiella tarda* and protective immune modulation in zebrafish larvae. *Fish Shellfish Immunol.* **2017**, *60*, 72–77. [CrossRef] [PubMed]
119. Ringø, E.; Olsen, R.E.; Gifstad, T.Ø.; Dalmo, R.A.; Amlund, H.; Hemre, G.-I.; Bakke, A.M. Prebiotics in aquaculture: A review. *Aquacult. Nutr.* **2010**, *16*, 117–136. [CrossRef]
120. Reza, A.; Abdolmajid, H.; Abbas, M.; Abdolmohammad, A.K. Effect of dietary prebiotic inulin on growth performance, intestinal microflora, body composition and hematological parameters of juvenile beluga, *Huso huso* (Linnaeus, 1758). *J. World Aquacult. Soc.* **2009**, *40*, 771–779. [CrossRef]
121. Mouriño, J.L.P.; Do Nascimento Vieira, F.; Jatobá, A.B.; Da Silva, B.C.; Jesus, G.F.A.; Seiffert, W.Q.; Martins, M.L. Effect of dietary supplementation of inulin and *W. cibaria* on haemato-immunological parameters of hybrid surubim (*Pseudoplatystoma* sp.). *Aquacult. Nutr.* **2012**, *18*, 73–80. [CrossRef]
122. Mahious, A.S.; Gatesoupe, F.J.; Hervi, M.; Metailler, R.; Ollevier, F. Effect of dietary inulin and oligosaccharides as prebiotics for weaning turbot, *Psetta maxima* (Linnaeus, C. 1758). *Aquacult. Int.* **2006**, *14*, 219. [CrossRef]
123. Geraylou, Z.; Souffreau, C.; Rurangwa, E.; Maes, G.E.; Spanier, K.I.; Courtin, C.M.; Delcour, J.A.; Buyse, J.; Ollevier, F. Prebiotic effects of arabinoxylan oligosaccharides on juvenile Siberian sturgeon (*Acipenser baerii*) with emphasis on the modulation of the gut microbiota using 454 pyrosequencing. *FEMS Microbiol. Ecol.* **2013**, *86*, 357–371. [CrossRef]
124. Geraylou, Z.; Souffreau, C.; Rurangwa, E.; D'Hondt, S.; Callewaert, L.; Courtin, C.M.; Delcour, J.A.; Buyse, J.; Ollevier, F. Effects of arabinoxylan-oligosaccharides (AXOS) on juvenile Siberian sturgeon (*Acipenser baerii*) performance, immune responses and gastrointestinal microbial community. *Fish Shellfish Immunol.* **2012**, *33*, 718–724. [CrossRef] [PubMed]
125. Grootaert, C.; Delcour, J.A.; Courtin, C.M.; Broekaert, W.F.; Verstraete, W.; Van de Wiele, T. Microbial metabolism and prebiotic potency of arabinoxylan oligosaccharides in the human intestine. *Trends Food Sci. Technol.* **2007**, *18*, 64–71. [CrossRef]
126. Parrillo, L.; Coccia, E.; Volpe, M.G.; Siano, F.; Pagliarulo, C.; Scioscia, E.; Varricchio, E.; Safari, O.; Eroldogan, T.; Paolucci, M. Olive mill wastewater-enriched diet positively affects growth, oxidative and immune status and intestinal microbiota in the crayfish, *Astacus leptodactylus*. *Aquaculture* **2017**, *473*, 161–168. [CrossRef]
127. Cicerale, S.; Lucas, L.; Keast, R. Biological activities of phenolic compounds present in virgin olive oil. *Int. J. Mol. Sci.* **2010**, *11*, 458–479. [CrossRef]
128. Ngugi, C.C.; Oyoo-Okoth, E.; Muchiri, M. Effects of dietary levels of essential oil (EO) extract from bitter lemon (*Citrus limon*) fruit peels on growth, biochemical, haemato-immunological parameters and disease resistance in Juvenile *Labeo victorianus* fingerlings challenged with *Aeromonas hydrophila*. *Aquacult. Res.* **2017**, *48*, 2253–2265. [CrossRef]

MDPI AG
Grosspeteranlage 5
4052 Basel
Switzerland
Tel.: +41 61 683 77 34

Foods Editorial Office
E-mail: foods@mdpi.com
www.mdpi.com/journal/foods